GPS
for Everyone:

You are Here

To Pat,
Colleague & friend,
with warm regards,

Pratap
10 Nov '16

Pratap Misra

Ganga-Jamuna Press

GPS abbr. noun

A global system of U.S. navigational satellites developed to provide precise positional and velocity data and global time synchronization for air, sea, and land travel (http://dictionary.reference.com)

A U.S.-owned utility that provides users with positioning, navigation, and timing services (http://gps.gov)

ISBN-10: 0-9709544-3-3
ISBN-13: 978-0-9709544-3-5
Copyright © 2016 Pratap Misra

Book Design by Jim Bisakowski, http://bookdesign.ca

Cover art: Getty Images

Ganga-Jamuna Press

P.O. Box 633

Lincoln, Massachusetts 01773

PratapMisra@ieee.org

http://www.GPSforEveryone.com

For Sita, Usha,
Ravindra, Shachindra, Mohanji

Contents

About This Book

GPS seems to have come out of nowhere.

There was no progression like eight-track tape to cassette to CD to MP3 player. One day we were driving around clueless of where we were, struggling with road-maps bought at gas stations that couldn't be folded back neatly once opened, and—suddenly—there was an amiable female voice coming out of the dashboard offering directions to our destinations and showing no signs of impatience when we made wrong turns.

Life before GPS

Cabbie's Question

This book is an attempt to answer a question a taxi driver once asked me: How does this little box sitting on the dashboard know I missed a turn back there?

Having spent much of my professional career on GPS, I felt I should answer the question. But there was no time and there was no blackboard. It took me a while, but I now have a short answer to the cabbie's question, which appears below, and a fuller one that's presented in the chapters that follow.

What is GPS?

Before we discuss how GPS works, let's talk a little about what GPS is.

GPS (full name: Navstar Global Positioning System) is a satellite-based, global radio-navigation system. That means it exploits the properties of radio signals, which are broadcast from satellites and cover the earth. GPS can be thought of as made up of three elements, called segments, which make up its infrastructure: Space Segment, Control Segment, and User Segment.

The Space Segment comprises a constellation of about two dozen satellites in inclined, circular, 12-hour orbits revolving around the earth. The altitude of the satellites is about 20,000 kilometers, or about three earth radii. Each satellite continuously broadcasts signals, much like an AM or FM transmitter.

The Control Segment comprises about a dozen sites scattered around the world, which monitor the satellites and their signals continuously and periodically upload new data for the satellites to broadcast. The U.S. Government is responsible for both these segments, whose development began in the early 1970s under the auspices of the Department of Defense (DoD), primarily for the benefit of the U.S. military.

The User Segment for our purposes is the receiver bazaar driven mostly by the market for such devices and their many applications, which took off in the mid-1990s and exploded with the advent of smartphones in mid-2000s.

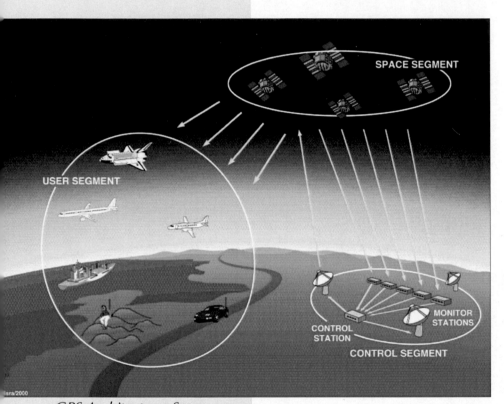

GPS Architecture: Space Segment, Control Segment, and User Segment

Back to the Cabbie's Question

So, here's the short answer to the cabbie's question.

The "little box on the dashboard," often referred to as GPS, is in fact a GPS receiver. The distinction between the source of the signals (GPS) and the receiver (the little box) is an important one to keep in mind. Actually, the little box is both a radio receiver and a digital computer with a display screen. The radio receiver part is designed to 'listen" to the signals transmitted by GPS satellites in the sky and to process them to determine (i) your range from each of the satellites, and (ii) the positions of the satellites. Given this information, the computer calculates your position, as in coordinates (x, y, z) or latitude, longitude, and altitude. That's all GPS can do.

If the receiver had a good view of the sky, your position estimate would be good to a couple of meters. GPS doesn't know the name of the street you are on, or how to get to 123 Main Street, if that's your destination.

In order to make the information provided by GPS useful to you in navigating to your destination, the computer in your GPS receiver takes the next step of displaying your position on a street map stored in its memory and showing what's around you. The part about directions to your destination and the sexy voice with different languages

and accents is all done by the people who built the digital computer part, and the reason you bought the little box. By matching your current position provided by GPS with the street map and the directions to your destination it had previously generated, the computer knows that you have gone past the intersection where you should have turned. That's about it.

The little box in our example above only used a *static* data base of street map stored in its memory. A GPS receiver-on-a-chip in a smartphone, with wireless communications and Internet access can give you *dynamic* information such as there is a traffic tie up a couple of miles down the road you are on; or a friend of yours is currently at the restaurant where you are getting ready to order your dinner; or your girlfriend just arrived at the apartment of an old flame. (There may be legal issues associated with why and how you received the last message, but we won't get into them here.)

So, if the little box gave you bad directions, don't blame GPS. The same if you made a mistake in entering information into the little box, as a Swedish couple apparently did a few summers ago. Heading from Venice to the Isle of Capri, they entered "Carpi" instead, and drove 400 miles to a grimy town in Italy's industrial north.

The Second Most Useful Gift of the DoD

There appears to be a consensus that GPS is the DoD's second most useful gift to the civil world, the Internet being the first, of course. But there is no shame in being second to the Net, which now plays a vital role in the newer applications of GPS as we go from *Locate Me!* to *Locate my Friends!*, *Locate my child!*, *Locate my Dog!* And, sadly, there is a societal need to locate and track thousands of men and women to keep them from entering individually tailored areas forbidden to them. But GPS is not a freeloader on the Internet. It reciprocates by providing precise time (accurate to within a fraction of a millionth of a second) to the Internet nodes to keep the data flowing smoothly.

The excitement of the discovery of GPS by ordinary users was brought home to me a dozen years ago when I entered a hospital for heart-bypass surgery.

So, there I am lying on a gurney being prepared for surgery by an anesthesiologist. It is early in the day and the anesthesiologist looks sleepy, seeming to make an effort at small talk. Now, I am not in a chatty mood either until, in response to a question about my profession, I say something about GPS and, suddenly, the anesthesiologist is fully alert. Turns out he is an instrument-rated pilot and GPS has changed his life. He is holding in his hand this syringe with stuff in it that would knock me out, but he is now in no hurry. He wants to talk about GPS-guided precision approaches and glide paths and decision heights, and it goes on for a while.

Apple Computer photo

Smartphone is an ideal platform for GPS
A smartphone has everything: access to data bases for things that don't change (street maps, business locations); Internet for things that change (traffic, road closures, sales); and instant communications.

During surgery, he tells the cardiologist that the guy on the cutting table knows a little about GPS. Two days later, the cardiologist visits me in my hospital room. I am anxious to talk with him about this foot-long cut in my chest, but he wants to talk about—GPS! What a thrill. Turns out, he is an avid sailor and GPS has changed his life.

This book is an expanded version of what I would have told these two excited physicians if I were in a shape to do so at the time.

Now Everyone is a Navigator

Navigators used to be professionals. They took measurements using specialized instruments and had the skills to calculate their positions from the measurements using disciplines like spherical trigonometry. And it wasn't that long ago; we are talking about mid-20th century.

Navigating in the air was different from navigating over water or land. Each required a different technology and different instruments. Transport airplanes had a navigator, who occasionally stuck his head out of the cockpit to take star sightings with a sextant. Closer to landing, they relied on an elaborate ground-based infrastructure of radio signals to guide them to the runway. Seagoing captains sailed on the open ocean with celestial measurements, and later Loran C, and had buoys and lighthouses to guide them into harbor.

Now, all anyone needs is GPS.

<aside>
GPS Didn't Come out of Nowhere

It just seems that way.

The basic ideas behind GPS have been around for centuries, but their implementation had to wait until the last-third of the 20th century when the required technologies matured and came together.
</aside>

What GPS can't Do

Now that we have a rough idea of what GPS is and how it works, lets dispel some common misconceptions. We won't spend any time on them here. That's what the rest of the book is for.

- GPS doesn't know your position; that's what your receiver is for.
- GPS doesn't know you live on a one-way street, or that you live.
- GPS has no clue about where you want to go, or how to get there.
- GPS can't track you.
  If someone wants to track you in real time, they'd need to stick a GPS receiver to you and have its position transmitted to them on a radio link. It'd be easier just to save your time-stamped positions inside the receiver and download to a computer at the end of the day to see where you have been.
- GPS can't be used everywhere.
  You must be able to see enough of the sky to have at least four satellites 'in view.' Over an ocean, you can get by with three. GPS radio signals are

not just weak but pathetic, and easily blocked by buildings, even trees.

- With parts worth $25 from Radio Shack, you can jam GPS for 100 kilometers. A hacker with $10,000 to spare can make you think you are someplace else, if your GPS receiver didn't anticipate such antics.

Economic Value of GPS

The economic value of GPS is approaching $100 billion a year.

Just imagine the savings from millions of motorists not driving around while lost, or thousands of delivery trucks optimizing their routes to avoid traffic jams and unnecessary travel. Transportation (land, air, and marine) aside, applications in construction, agriculture, mining, scientific studies, commerce (location-based services), and asset tracking have turned GPS into a powerful economic force.

The role of GPS in social media, though large, is still evolving. And we haven't even mentioned the military applications, the *raison d'être* of GPS.

No wonder Russia wants a GPS of its own. And so do Europe, China, Japan, and India.

Outline of the Book

For the long answer to the Cabbie's question, we'll expand on what we have outlined above and look back to see how we got here.

GPS didn't come out of nowhere. It just seems that way. The basic ideas have been around for centuries, but their implementation had to wait until the last-third of the 20th century when the required technologies matured and came together.

GPS gives your position within a couple of meters. It gives even sharper estimates of velocity and time. The timing part may well be a stealth service. If we lose GPS for a day, the loss of time may hurt us the most as the trading on Wall Street, bank transactions at an ATM, digital communication networks, and the power grid struggle to cope with the loss of a precise and universal timing service they have come to rely on.

In the rest of this book, we'll talk about:

- basic principles behind the operation of GPS,
- technologies central to the realization of GPS,
- GPS signals and how a receiver processes them to determine position, velocity, and time,
- the many applications, civil and military, and
- the story of how GPS came to be developed and the men (yes, they were all men—we are talking about the 1960s and 70s), who contributed vital

How Much Does GPS Cost?

It depends.

If you are an American, you paid for GPS with your taxes and you own it. You spent about $20 billion on its design, development, test, and deployment spread out over 22 years, from 1973 to 1995. It costs you about $1 billion each year to operate and maintain GPS.

Believe me, you got a super deal.

If you are not an American, you got an even better deal—you got GPS for free.

ideas to its design and organizational savvy to its realization.

We don't much dwell on the history of navigation tools and techniques. Much has been written already about the desperate search over centuries to find longitude at sea, to which we have nothing to add.

Prerequisites for This Book

The only prerequisite for this book is curiosity about a technology that has insinuated itself into our daily lives over the past few years.

A reader who remembers some of what he or she learned in high school mathematics and physics courses would be able to follow much of the discourse, even in chapters where we attempt a serious discussion of the signals and receivers. A student enrolled in high school AP physics class would understand everything. Some equations are included as convenient shorthand, but these may be skipped by a reader so inclined without loss of continuity.

Acknowledgment of Debts

I was fortunate to have stumbled into the field of navigation satellites 25 years ago and to have found generous mentors and colleagues, several of whom have assisted me with this book. Per Enge, a longtime friend and collaborator, will recognize the imprint of his ideas in a number of passages. I am indebted to Simon Banville, Ray Filler, Gaylord Green, Gérard Lachapelle, Dian Pekin, and my daughter Tara, who reviewed an early draft and offered ideas for improvement. My debt to Richard Langley is even greater: Richard's painstaking review of the final draft has saved me from embarrassment on matters both technical and stylistic.

I am grateful to Ray LaFrey and Vince Orlando, former colleagues at MIT Lincoln Laboratory, for giving me an opportunity in 1989 to learn about GLONASS, the Soviet answer to GPS. More recently, I have been able to count on the support of my colleagues Bill Messner, Jason Rife, Chris Rogers, and Rob White of Tufts University; Ray Filler and Steven Ganop of Penn State/ARL; and Paul Manz of PEO Ammunition, U.S. Army. I thank Lisa Beaty for creating a nurturing place for GPS engineers at the Institute of Navigation. Finally, a debt of gratitude is owed to Brad Parkinson, whom we call the Architect and Prime Mover of GPS. More on that in a later chapter.

Pratap Misra
1 July 2016

Want to Learn More about GPS?

If you want to dig deeper into the structure of the GPS radio signals and how they are processed by a receiver to determine your position, try an engineering textbook I coauthored a few years ago with Professor Per Enge of Stanford University (http://www.GPStextbook.com).

GLOBAL POSITIONING SYSTEM
Signals, Measurements, and Performance
Second Edition
Pratap Misra
Per Enge

Where are You?

"I am waiting in front of the Minuteman statue in Lexington Center. *Where* are you?"

That's how we locate ourselves and others in everyday life: relative to known landmarks. We name streets, assign numbers to buildings, and draw street maps of towns. An answer like "I am at 123 Main Street" tells your friends exactly where you are. It works pretty well for us much of the time.

But this approach wouldn't work if there were no identifiable landmarks, or streets had no names, or worse, there were no streets. What if a place has been hit by a disaster—an earthquake, hurricane, or tsunami—wiping out landmarks, tearing road signs, blocking streets with debris, and filling the place with smoke. How are the emergency personnel to navigate?

How to assign addresses in a community we call slum, basti, or favela that developed helter-skelter without the benefit of roads, urban planning, and much else, where one tarp-covered dwelling looks like another. What if you were in a boat in an open ocean, or crossing a desert, or flying in an airplane above the clouds, or hiking in a wilderness area?

If you can't associate your position with a landmark, you need to describe it in a way that would make sense to someone in the next town, or another continent. That means developing a language of location. Many such regional and local languages were developed over the years, but GPS needed a language

Waiting at the Minuteman statue (photo by John Scott Parker)

that would be understood all over the world—much like English in modern communications. There is a difference though. We are not talking about just anything. We want a language simply to characterize a position unambiguously and precisely. And that means a language with limited, simple vocabulary and tight grammar. We'll say more.

There are two parts to what we are after broadly. First, we want to know in an unambiguous way where we are and, second, how to chart a course to our destination. The art and science behind how we answer these two questions is what we'd call navigation.

What street address? (Africanewspost.com photo)

What landmark? (USAF photo)

Human Navigation Sense

Navigation, unlike texting or googling, is not a new human activity.

Our oldest ancestors would have had to find ways to navigate to a fishing spot or hunting ground, and navigate back to their caves using visual clues and landmarks. You can imagine a cave elder instructing a young kinsman: Get to high ground, locate the sprawling banyan tree in the distance and walk toward it; climb up the banyan tree and locate on the horizon a big rock with a pointy top; you'll come upon the best fishing spot in the area as you walk toward that rock.

When it comes to innate navigation sense, we humans are a pathetic lot.

Butterflies find their way to specific wintering sites thousands of kilometers away where they have never been. Swallows find their way back to San Juan de Capistrano from who-knows-where, arriving on a specific date each year. The slight, silvery California grunion emerge from the Pacific at the sandy beaches at night to spawn and dance with predictable regularity twice a month in early spring with high tide under a full or new moon. But we may have trouble in daylight telling which way is north.

What we have going for us, though, is a sense of curiosity and capacity to learn and invent. Over centuries, we learned to use the sun, stars, and the earth's magnetic field, apparently as the birds do, and developed navigation instruments: magnetic compass, sextant, time-keeping devices, inertial instruments, and radio-navigation systems Loran, Omega, Transit, and—a crowning achievement—GPS.

Monarch butterfly, migration paths, and overwintering in Central Mexico (Wikimedia Commons)

Every Point Has a Unique Geographical Address

Every point on the earth's surface has a unique location and could be assigned a unique geographical address if we had a fine, uniform grid covering the earth's surface.

This would have been easy if the earth were flat like a tabletop. We could then define an origin and two lines at right angles as x-axis and y-axis. That's all we need in order to describe the position of an ant moving on the table in terms of its x- and y-coordinates. Defining a uniform grid wouldn't have been much harder if the earth were spherical like a beach ball. We'd place the origin at the center and define an equatorial plane and an axis perpendicular to it. We could then describe every point on the surface of the ball in terms of two angles we could call latitude and longitude.

Our job would get much harder if asked to devise a grid or a coordinate frame to express the positions of points on the surface of an irregular rock you might find on the side of the road. It's no longer clear where to put the origin and how to draw the axes. The earth falls somewhere between a beach ball and irregular, inhomogeneous rock.

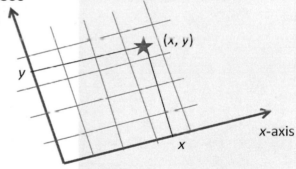

A two-dimensional Cartesian coordinate frame

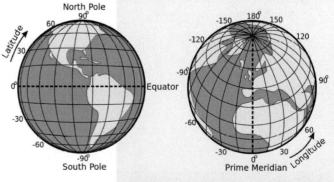

Latitude and longitude defined on a spherical surface
(*www.geographyalltheway.com*)

In a large, independent effort that anticipated GPS, the DoD has developed and refined since the 1950s a global coordinate frame that's now good to centimeter level. We'll define what that means and say more later. For now, suffice to say that GPS gives your position in this coordinate frame virtually instantaneously with an error that's typically no more than a couple of meters.

We have created maps of the stationary objects, like roads, buildings, and monuments relative to this GPS grid and can show positions of moving objects, like our cars, on such maps.

Insofar as assigning a geographical address goes, we are not limited to the earth's surface. Any point in space, whether on, above, or below the earth's surface, has a unique geographical address. In order for GPS to determine these coordinates, the requirement is simple: GPS signals must reach the point of interest. There are some other caveats, but we don't want to get tied down by small print so early in the book.

My Coordinates Are:

So, you can tell your impatient friend waiting at the Minuteman statue in Lexington, Massachusetts, USA, your position exactly. Your (x, y, z) coordinates as provided by GPS are (in meters):

$$\begin{bmatrix} x \\ y \\ z \end{bmatrix} = \begin{bmatrix} 1\,510\,885 \\ -4\,463\,460 \\ 4\,283\,907 \end{bmatrix}$$

But, like most GPS users, you don't want to get tangled up in numbers, and nor does your friend. We have computers for that.

You can transmit your coordinates simply by pressing a couple of buttons on your GPS-capable cellphone so that your position appears as a bright dot on a street map display on her cellphone, and she'd know that you are about half-way down Wood Street and no more than five minutes away.

The playing field is finally level and no migratory animal species is going to look down its nose at us.

Where is Basher 52?

Most of us look to GPS for directions to a destination and perhaps ways around traffic jams along the way. It's now an everyday thing, nothing to get too excited over. But sometimes GPS can save a life and the story can become front-page news and lift the spirits of millions. Let's start with one such early story.

The public announcement came on 17 July 1995 that GPS was 'operational,' meaning the satellite constellation was fully populated and the system was ready to be used for real. The surveyors and geophysicists already had been using it for a dozen years. The U.S. military had recognized its extraordinary value in combat in the first Gulf War in 1991 and had moved quickly to incorporate it in operations. Inexpensive Garmins and TomToms were a few years away. The World Wide Web was just beginning to catch on. The cellphones were just that—telephones, and clunky-looking at that. The idea of an iPhone with or without GPS hadn't yet occurred to anybody at Apple.

Back Story

Our story is set in the former Yugoslavia. Following the breakup of the Soviet Union and fall of the Berlin Wall, Yugoslavia started cracking up under the weight of ethnic divisions (Serbs, Croats, Bosniaks, Slovenes, Macedonians), religious divisions (Eastern Orthodox Christians, Roman Catholics, Muslims), and mutual grievances accumulated over centuries. Slovenia and Croatia, overwhelmingly populated by a homogeneous group, broke away in 1991. Bosnia, with a Muslim majority but significant Serb and Croat population, turned into a slaughterhouse.

Socialist Federal Republic of Yugoslavia as of January 1991 (from U.N. website http://www.icty.org/sid/321)

Slobodan Milosevic, President of Serbia, wouldn't allow Muslims in Bosnia the right to govern themselves and the Serb minority. He sponsored a systematic campaign of ethnic cleansing. It was said that the Bosnian Muslims appeared too European to the Saudi Arabians and too Muslim to the

Europeans, and the world looked the other way for a long time as they were slaughtered between 1992 and 1995.

Finally, NATO intervened with air strikes, forcing Milosevic to negotiate. A peace accord was signed in Dayton, Ohio, in December 1995, and the Republic of Bosnia-Herzegovina became a reality. Milosevic was tried at the International Court of Justice in The Hague for crimes against humanity, but died in his cell in 2006 before the trial was complete. His brutal henchmen, Radovan Karadic and Ratko Mladic, the overseers of the blood bath in Bosnia, went into hiding in Serbia. Karadic was apprehended in 2008 in Belgrade. Mladic evaded justice until 2011. Both are now on trial in The Hague to answer for the siege of Sarajevo and Srebrenica massacre.

That's all for the back story. Let's get to the part about GPS.

Pilot is Down

On June 2, 1995, Capt. Scott F. O'Grady, a U.S. Air Force pilot, call sign Basher 52 (pronounced Basher five-two), was patrolling the no-fly zone over Bosnia in an F-16 at 26,000 feet when his plane was struck by a Serbian ground-to-air missile and broke up in mid-air. Capt. O'Grady managed to eject himself from the disintegrating plane and parachuted—not unnoticed by the local Serb militia—into the countryside of Bosnia. Unfortunately, his wingman didn't see the parachute emerging from the wreckage, leading his comrades to fear the worst.

Capt. Scott F. O'Grady and F-16 fighter aircraft
(U.S. Air Force photos)

The 29-year-old pilot, wearing only a T-shirt under his flight suit, managed to evade the Serbian search parties looking for him, sleeping by day under camouflage netting and moving at night, and surviving on grass and insects, putting into practice lessons from his 17-day USAF training in Survival, Evasion, Resistance, and Escape (SERE, pronounced seer).

Six days later, O'Grady heard his call sign on his survival radio.

> Basher 52, this is Basher 11. Basher 52, this is Basher 11. Are you up on this frequency?

> This Is Basher 52 … I am alive and I need help.

> Say again Basher 52. You are weak and unreadable.
> This Is Basher 11.

Yes, weak and unreadable to Capt. T. O. Hanford (Basher 11) in his F-16 about 100 kilometers away from the pine forest where Capt. O'Grady was hiding. It was 2:08 a.m. on June 8. After verifying his identity (what was he called in high school when he got drunk?), the pilot passed the information to the Air Force NATO Command. By 3:00 a.m., a decision had been made to mount a rescue operation and the primary rescue team from the 24th Marine Expeditionary Unit under the command of Col. Marty Berndt on USS Kearsarge in the Adriatic Sea swung into action.

Search-and-Rescue (Minus the Search)

At 5:05 a.m., two CH-53 helicopters carrying 43 Marines—riflemen, assault climbers, medics, a communications team, and an interpreter—lifted off the flight deck followed by two AH-1W Cobra gunships and two AV-8 Harrier jump jets. Soon an armada of 40 aircraft, jammers and fighters, and an AWACS air traffic control plane had assembled in the air over the Adriatic Sea and western Bosnia. The rescue operation got underway at 6:12 a.m.

As the day was breaking, the two Marine helicopters landed in a clearing in the pine forest near the city of Banja Luca. The Marines quickly jumped out to secure the perimeter. Capt. O'Grady ran out of the woods and jumped in a helicopter as small arms fire was heard in the background. A sergeant offered Capt. O'Grady a drink, then put away his metal water bottle just in time for it to deflect a Serb bullet which had penetrated the fuselage. The helicopters were quickly airborne. The rescue had been completed within five hours of the radio contact with Capt. O'Grady. There were no casualties.

After six days of insects and grass, Capt. O'Grady feasted on the military's MREs (meals ready to eat) amid his rescuers aboard the CH-53. By 7:30 a.m. the team had landed on USS Kearsarge with their now-famous rescued pilot. The word of successful rescue was quickly flashed to the White House. President Clinton stepped out on a balcony along with Anthony Lake, his National Security Advisor, to celebrate with a victory cigar.

Capt. O'Grady was hailed as a hero, but he said the real heroes were the Marines who had rescued him. It took a while to

Capt. O'Grady at the White House
(White House photo)

Trimble Flightmate GPS was a part of Capt. O'Grady's survival kit (Credit: Trimble Navigation)

Combat Survivor/Evader Locator (CSEL) is an advanced version of what Capt. O'Grady carried in his survival kit (Credit: The Boeing Co.)

recognize, but another quiet hero of the successful operation was a system few had heard off—GPS.

Flightmate GPS

Among items in Capt. O'Grady's survival vest was a handheld GPS receiver, a Flightmate made by Trimble, a company based in Sunnyvale, California, a pioneer in GPS receivers. Other items were: survival radio, distress signals, mirror (for signals, not makeup), first-aid kit, compass, face paint for camouflage, and tourniquet. The pilot knew his position coordinates. The search-and-rescue party didn't have to search for him. Total time spent by the Marine helicopters on the ground was less than three minutes. It was a textbook rescue.

The Trimble Flightmate was a civil GPS receiver. In 1995, it was the U.S. policy to degrade the GPS signals available to civil users in order to reduce their positioning accuracy to 100 meters, but that was good enough for the rescue mission. (This policy, called Selective Availability, didn't sit well with the civil users and was discontinued in 2000.) A military receiver could have given Capt. O'Grady's position within several meters. But the civil receivers as a rule are light, sleek, and easy to use; the military receivers, just the opposite. A military receiver of 1995 wouldn't have fit in the survival vest.

The newer survival radios are much more capable. Capt. O'Grady was required to read his position from the GPS receiver display and speak the numbers into his AN/PRC-120, a line-of-sight UHF-VHF radio, to let his comrades know on an unsecured radio link open to eavesdroppers. The newest version, called Combat Survivor/Evader Locator (CSEL) has an embedded military GPS receiver and, when switched on, automatically sends encrypted signals, including position, directly to a search-and-rescue center via communication satellite links using low-probability-of-detection signals, meaning signals that are so weak that they wouldn't register on the enemy's frequency scanner. (We'll have more to say about these spread spectrum signals later when we discuss the GPS signals.)

If Capt. O'Grady had a CSEL in his survival vest, he would have been rescued during his first night in the pine forest of Bosnia.

What a Simple Idea!

GPS is based on an idea that has been around forever and you know it already.

Where did the Lightning Strike?

It works like this: When you see a flash of lightning, you start counting to keep time in seconds—one Mississippi, two Mississippi, and so on. If you got to 'three Mississippi' before you heard the thunder, the lightning was about a kilometer away. The idea is very simple. The flash of lightning and the thunder occurred simultaneously. Light travels so fast as to reach you virtually instantaneously; sound travels much slower in the air—about a kilometer in three seconds.

We can extend this idea to tell *where* the lightning struck, not just how far away. To do this, you have to enlist a friend who lives a few kilometers away to count Mississippis with you, to obtain estimates of distances to the lightning from your house and hers. Now you pull out a town map, locate both houses on it, and draw circles to scale centered at the two houses with radius of each equal to the corresponding measured distance. The lightning struck at the point where the two circles intersect on the map.

Speed of light: 300 000 000 meters per second

Speed of sound: ~ 350 meters per second

Your friend's house

Your house

Actually, the two circles intersect at two points and we have an ambiguity. Each of these points is consistent with the measurements. You now have to find a way to eliminate one of the points with additional information: for example, you observed that the lightning was to the north of your house. Otherwise, you would have to find a second friend willing to humor you, and you'd have three circles to draw.

Before we leave thunder and lightning behind, let's take one more crack at this analogy to determine *your* position.

MacGyver-esque Positioning

So, you are on a day hike up a mountain when a thunderstorm develops unexpectedly. You scramble off the trail in search of shelter and eventually find one under a rock outcropping. The thunderstorm shows no sign of letting up and it's beginning to get dark. You figure you wouldn't be able to find the trail in the dark and need help getting down. You have a cellphone and can call 911, but what are you going to tell them about where you are? What would MacGyver do?

You look around and see lightning striking repeatedly two tall church spires—one in front of you and one to your right. You whip out a map of the area which shows the locations of the two churches. You estimate the distances to the two churches by estimating the time it takes for the thunder to reach you from each and using a compass—you always keep one in your backpack—draw two circles centered at the churches as in our previous example. Your position on the map is given by the point where the two circles intersect.

A call on the cellphone to tell the rescuers your position, and happy ending.

As a postscript to this story we should add that as Extended 911 service (E911) spreads throughout the country, all you would have to do is to call 911 on the cellphone. The smartphone would have determined your position using the built-in GPS receiver and communicated it automatically to the rescue unit. You don't even have to worry about packing a compass in your hiking bag.

We have removed all evidence of murky weather to illustrate the principle of trilateration.

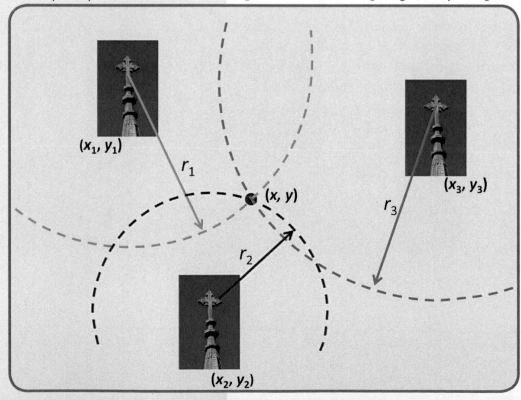

(x_1, y_1)

r_1

(x, y)

r_2

(x_2, y_2)

(x_3, y_3)

r_3

Trilateration

Let's revisit briefly our geometric construction above for determining an unknown position in order to make a couple of important points. The construction involved drawing circles which were centered on objects whose positions were known and whose radii were the measured ranges.

Clearly, if you knew what we have called 'known' positions perfectly, and had perfect range measurements, you'd have determined the unknown position without error. In real life, there is no such thing as perfect knowledge or perfect measurements. The known positions would have errors in them, so would the range measurements. Clearly, the smaller the errors, the better the position estimate.

We'll make a case on an intuitive basis from the figure on the opposite page that if you want your position with meter-level accuracy, the positions of the spires have to known with meter-level accuracy and the ranges have to be measured with meter-level accuracy. While we had no such expectation in the examples cited above, we'll repeat this *mantra* as we get deeper into how GPS works.

How many range measurements did we need? In the simple examples above dealing with the lightning, we needed at least three for an unambiguous answer. The number remains three as we go from two-dimensional positioning to three-dimensional positioning. The basic idea is the same, but circles get replaced by spheres.

Determining the position of a point by measuring its distance from points at known locations is called *trilateration*. You have probably heard of its better-known cousin, *triangulation*, a technique for determining the position of a point by measuring angles to it from two locations whose position coordinates are known. The point could then be 'fixed' as the third point of a triangle with one known side and two known angles.

Triangulation was the standard technique of land surveying for hundreds of years using an angle-measuring apparatus and iron chains to measure lengths. The technology for measuring even moderate distances accurately did not exist until the middle of the 20th century when laser range finders became available, giving surveyors relief from hauling their bulky chains.

In order to find the coordinates of a point in the U.S. just 50 years ago, you would have had to locate in the neighborhood two *survey markers*—points whose positions had previously been determined by the U.S. National Geodetic Survey, and imprinted on brass disks affixed to the ground. You'd then have stationed a theodolite mounted on a tripod atop each point and measured the angle to the point to be located, which had to be visible from the two survey markers—a serious complication.

That's all for the look backward.

Survey marker (NOAA photo)

Surveying Kodiak Island, Alaska, by triangulation in 1929 (NOAA photo)

Conceptual Design of a Radio-Positioning System

Let's now try to extend the scheme we contrived to rescue the lost hiker into something usable by many to locate themselves over a large area. We'll deal with broad concepts for now and not worry about the engineering details of implementation. There will be time for that.

What were the main elements of our scheme? We had church spires as survey markers; we had lightning and thunder to measure ranges to these markers; and we had a town map on which to represent the positions of the spires and the hiker.

What would it take to build a positioning system usable over a wide area based on this idea? First, we'd have to raise the spires really, really high. Second, we'd need a capability to call in lightening as needed. Third, we'd have to build a map showing the positions of the spires and the users of our system.

Yes, that would work. What shall we call our system? Well, let's finish the design first.

Okay, we'd raise our church spires by thousands of kilometers into space, replacing them with satellites. The satellites would be our survey markers. They'd be going around the earth at several kilometers per second, and it'd be a challenge to know their positions at any instant within meters, but it may be doable.

We'll use radio signals to extend the idea of range measurement with lightning and thunder. The satellites would transmit radio signals and we'd measure how long it took them to reach you. Radio signals travel at the speed of light—one meter in about 3-billionths of a second. We'd need very precise and accurate clocks. An error of a thousandth of a second will translate into an error of 300 kilometers in range, possibly placing us in a wrong state, perhaps even a wrong country.

Going from a village map to a global, three-dimensional grid would be tricky, but it's just a matter of scale. Nobody said it was going to be easy.

We are basically done. We'll refine our ideas as we go from conceptual design to the real thing. But, we've got satellites; we've got radio signals; we've got accurate clocks; and we've got a global grid in which to express positions. That's it!

So, what shall we call our system? How about GPS?

We have now introduced the main ideas and created an outline of what GPS is and how it works. Before we fill in the outline, we have to introduce an important missing piece. The main thing missing from our account so far is that GPS is not the first navigation satellite system and it didn't have to start from scratch and invent everything.

Before GPS there was Transit, which became operational 30 years earlier. But unlike GPS, Transit wasn't based on trilateration; it wasn't continuous; it wasn't instantaneous; it didn't offer meter-level accuracy; and it didn't see mass use. Still, Transit was revolutionary, even magical.

Let's begin by invoking Arthur C. Clarke, who famously declared that "any sufficiently advanced technology is indistinguishable from magic." Say what you will about Clarke (or Sir Arthur, if you are an Anglophile), but there is no denying that he was a technological visionary. He seems to have been the first to see the potential of satellite-based relays for global radio communications, but conceptually that seems simpler than satellite-based navigation using radio waves. (We'll meet the inventor of satellite navigation in this chapter.) A 1956 letter of Clarke's has made the rounds on the Internet in recent years. It contains a throw-away line about how three geostationary satellites could make possible a "position-finding grid whereby anyone on earth could locate himself." We don't know exactly what Clarke had in mind and we are not going to parse that sentence too closely. Having tipped our hat to Clarke, we'll move on.

Arthur C. Clarke (1917-2008)
(ucsdnews.ucsd.edu)

GPS was declared ready for general use on 17 July 1995. Actually, in bureaucratic language of the U.S. Government's acquisition process, GPS was said to have reached *Final Operational Capability* on that date. To answer the question posed in the title of this chapter, GPS could perhaps have become operational 10 years earlier, if the U.S. Department of Defense had foreseen how it would turn out. But even our brightest didn't. That's true of the Internet, too. Who knew in the 1970s that ARPAnet, which connected a dozen universities whose professors were involved in DoD-sponsored research, would evolve into the Internet, now indispensable to our lives?

But let's get back to Transit and get the history of satellite-based global radio-navigation straight. The system had a long, clunky official name, but like everybody else we'll call it Transit. It was based on a brilliant invention that occurred in a flash as the first spacecraft marking the dawn of the space age was still in the sky and was realized with extraordinary speed. Transit went on to serve the U.S. Navy, and later the commercial fleet, until made obsolete by GPS.

Transit happened so fast that its storyline and cast of characters remained lean. That's not true of GPS. We tell the story of Transit briefly in this chapter and then talk about how GPS built on this success.

Space Age Begins: Sputnik I

The solar activity in 1957 was expected to reach a periodic peak. This activity would be reflected in the ionosphere, a shell of charged particles stretching from about 50 kilometers to 1000 kilometers above the earth, whose intensity would build up to an 11-year high. The geophysicists were excited about the prospects of observing the ionosphere from above for the first time with a 1.5-kilogram instrumented spacecraft the U.S. had planned to launch to commemorate what the scientists were calling the International Geophysical Year.

Then out of nowhere came news on 4 October 1957 that the Soviet Union had launched a shiny, 84-kilogram, beach ball-size sphere called Sputnik I into orbit. Sputnik in Russian literally meant a traveling companion, but that's what a satellite is. It was a Soviet gesture to commemorate the IGY. It was all about science.

But everybody knew it was really all about Cold War, fought over four decades during which the two superpowers engaged in menacing displays of superiority. If the Soviets had launch capability to hurl an 84-kilogram payload into space, they could do some scary stuff with ballistic missiles. The U.S. was unnerved. The Soviets announced gleefully that "the new socialist society turns even the most daring of man's dreams to reality." It was a crazy time. Imagine people talking like that, even in translation.

Sputnik 1 was a beach ball-size spacecraft with four giant whiskers

In a further setback, the first launch by the U.S. of their 1.5-kilogram spacecraft on 7 December 1957 failed and was derided as the 'Flopnik.' The U.S. achieved the first of its many successes on 31 January 1958 and went on to win the space race decisively. The unpleasant surprise of Sputnik led the U.S. to establish the Advanced Research Projects Agency (the previously cited ARPA, later renamed Defense Advanced Research Projects Agency, or DARPA) and NASA, which have served the nation well. We'll stay with Sputnik I for just a little longer to tell the story of the beginnings of satellite-based navigation.

Sputnik's Orbit

The Soviets were having too much fun with their toy in space to announce any details, like the specifics of its orbit. Sputnik carried two 1-watt transmitters and four 3-meter-long whip antennas that looked like giant whiskers. It transmitted very simple radio signals: pulses or 'beeps' at 20.005 megahertz and 40.002 megahertz. These beeps could be heard by anyone with a shortwave radio tunable to 20 megahertz or 40 megahertz.

The radios produced audible tones whose pitch changed as the satellite moved in the sky, ranging between 500 hertz and 1500 hertz. The amateur radio operators around the world were excited no end monitoring these beeps for 22 days until the batteries on Sputnik gave out. (Batteries were two-thirds of the 84-kilogram mass of the spacecraft.) You can replay the beeps even today on the Internet (http://history.nasa.gov/sputnik/).

Fortunately, among those monitoring the beeps were two young physicists at the Applied Physics Laboratory (APL) of the Johns Hopkins University: William H. Guier and George C. Weiffenbach. They were intrigued by the pattern of changes in the frequency of the signal from Sputnik as it rose above the horizon, climbed in the sky, declined in elevation, and set. The change in frequency was entirely expected due to the *Doppler Effect* we learn about in a high school physics class. It refers to the change in the apparent frequency of a signal received by an observer due to relative motion between the transmitter and receiver.

The Doppler Effect is familiar to anyone who has listened to a train whistle while standing on a side of the track: The pitch of the whistle starts out high and drops slowly as the train approaches and then drops sharply as the train passes and recedes. The classical explanation of this phenomenon is as follows. As the locomotive approaches, more cycles are received in a time interval than transmitted due to the shrinking distance between the transmitter and observer, and the pitch seems higher; the opposite is true as the locomotive recedes. Doppler shift, defined as the difference between the frequency of the received signal and the frequency at the source, is proportional to the rate of change in the distance between the transmitter and receiver.

Guier and Weiffenbach, being physicists, started recording and plotting the Doppler shifts for each pass of the satellite while it was in range. They needed very accurate timing signals, which they obtained from the National Bureau of Standard's radio station WWV 10 miles away. They were struck by how the pattern of Doppler shifts changed from one pass to the next. They knew that this pattern was tied to the change in relative velocity between Sputnik and their ground station.

The physicists also knew that orbital motion is regular and follows laws formulated by Kepler 350 years earlier. An orbit is fully characterized by a half-dozen parameters, appropriately called Keplerian parameters. It was simply a matter of fitting the Doppler data to the orbital model to estimate these parameters. Data from a single pass were enough to get decent estimates. Consistency of the estimates obtained from the different passes provided confidence that the estimates were good. Guier and Weiffenbach had invented Doppler tracking of satellites, which was more accurate than any method then available. We'll call them *Instigators of Satellite Navigation* for what followed.

Invention of Satellite Navigation

Frank T. McClure, center, whom we call Inventor of Satellite Navigation, flanked by colleagues William H. Guier (left) and George C. Weiffenbach, circa 1960   *(JHU/ APL photo)*

We now introduce the *Inventor of Satellite Navigation*: Frank T. McClure, the boss of Guier and Weiffenbach at APL. McClure made a brilliant deduction: If the orbit of a satellite can be identified from Doppler shifts in the signal measured from a single ground station at a *known position*, as Guier and Weiffenbach had shown, the inverse problem can be solved as well: We can determine the position of a ground station by measuring Doppler shifts in a signal from a spacecraft in a *known orbit*. Totally brilliant!

McClure's invention was timed perfectly. He knew that the Navy needed a new kind of navigation system to guide their nuclear-powered submarines carrying Polaris missiles. The fire control system had to know the submarine's position accurately in order to fire the missile at a target. The submarines, when submerged, used inertial navigation in which position errors build up over time. The best available inertial systems costing hundreds of thousands of dollars drifted at a rate of about 1 nautical mile per hour. To reset the system, the submarines surfaced periodically for star measurements, a lengthy and cumbersome process. The Navy needed a system that was global, accurate, quick, and passive (meaning, a user only listens). The 'passive' part was important: A submarine was not to radiate any signals so as not to be found out. A satellite-based Doppler navigation system would have been ideal, if one could be realized.

Unlike trilateration, Doppler positioning is not intuitive. A sidebar illustrates how you can determine your position in relation to a train track by measuring Doppler shifts in the pitch of the train whistle. And by that we mean position along the track and distance from the track.

Doppler Positioning

A train is traveling at a steady high speed on a straight and level track blowing its whistle. Three observers listen to the whistle as the train passes. We have set up our x- and y-axes along-track and cross-track as shown and would like to determine the position of each observer in this coordinate frame.

For the observer standing in the middle of the track (not recommended), the pitch starts out high and then drops sharply as the locomotive runs over him and recedes. An observer standing away from the track also experiences Doppler shift, but the change between the high and low pitch is smooth.

The farther away the observer is from the track, the lower the range rate, and the smaller and slower the change in pitch. The shape of the Doppler shift curve gives your distance from the track, or y-coordinate.

To determine the common x-coordinate, we need a watch and the train's schedule. Note the time when the pitch of the whistle was midway between the high and the low. Check the train's schedule to determine its position along the track.

That's essentially how Transit worked.

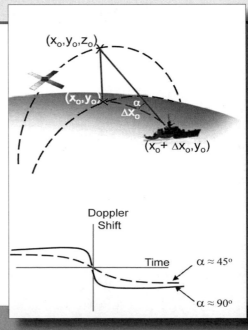

Genesis of a Navigation Satellites

So, as we were saying before being sidetracked by the train whistle, McClure had this idea of determining a position on earth by measuring Doppler shifts in signals transmitted from a spacecraft in a known orbit. We did point out that it was a brilliant idea, didn't we? The operational principle is simple: A ship determines its position, along-track and cross-track relative to the satellite orbit, actually its projection on the ocean surface, as with the train whistle, by measuring the Doppler shifts in a signal transmitted by the satellite.

Richard B. Kershner, Father of Satellite Navigation

McClure presented his idea to Richard B. Kershner, a colleague at APL, and together they quickly worked out the essential elements of the system. APL delivered a 50-page proposal to the Navy in the spring of 1959 and later that year to ARPA, the venturesome agency chartered the previous year specifically to take on high-risk, big-payoff projects. And there were risks aplenty.

In 1959, there was much to be learned about designing reliable, long-life spacecraft; launching them into space; keeping the antennas pointed toward the earth; and determining accurate orbits. Repair and replacement of a faulty part in a satellite would be a problem, and there were to be many new types of mechanical and electronic components operating in a whole new environment. In the first year of ARPA sponsorship, the Agency requested an analysis from an independent organization of mean-time-to-failure for a Transit satellite. The answer came back: 2 weeks.

Transit was a pioneering achievement of the APL, which was responsible for the whole thing from the initial concept developed in 1958 to the design of experimental satellites launched in 1961–1962, and the final system which became operational in 1964.

The Transit program was directed by Richard B. Kershner, whom we'll call *the Father of Satellite Navigation*.

Transit

Transit was realized with a half-dozen satellites in low-altitude (1100 kilometers), nearly circular, polar orbits. Each satellite broadcast signals at 150 megahertz and 400 megahertz with a total transmitted power of 1 watt. Only one satellite was in view at a time and a user waited up to 100 minutes between successive satellite passes to determine position.

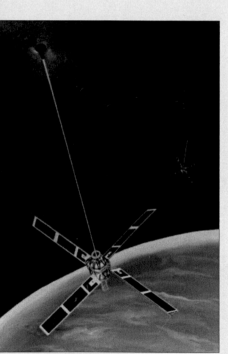

Transit satellite in space with a gravity boom for alignment
(APL/JHU graphic)

After a satellite came in view, a Transit receiver the size of a couple of filing cabinets, recorded continuously the Doppler shift of the received signal and the navigation message giving the satellite position during the satellite pass lasting about 10–20 minutes. These measurements were subsequently processed by digital computers the size of a couple of filing cabinets to compute the 2-dimensional position of a stationary or slow-moving ship. The computations took several minutes.

As with the train whistle, the Doppler shift is positive when the satellite is approaching, and negative when receding. How quickly it went from positive to negative depended on how high the satellite rose in the sky above the ship. The shape of the Doppler shift curve provided a measure of the user's distance from the satellite's ground track; the receiver clock and the satellite almanac determined the position along the ground track.

The typical positioning accuracy of Transit was a couple of hundred meters, an astonishing feat in the 1960s.

We now are ready to focus on GPS.

Requisite Technologies

Recall that for meter-level three-dimensional positioning continuously and instantaneously anywhere on the earth using trilateration, we said we needed:

- a constellation of satellites whose positions at an instant could be *predicted* with meter-level accuracy,

- a global coordinate system with meter-level accuracy—actually, *consistency*, as we'll see—to represent the positions of the satellites and users, and

- clocks aboard satellites that keep time accurate to within several billionths of a second over a day to measure ranges with meter-level accuracy.

Fortunately, the technologies of orbit prediction, global coordinate frames, and timekeeping matured and came together in the last third of the 20th century. We look into their development in this chapter, and then take another crack at describing the design of GPS in the next chapter.

6.1 Space-Based Platforms in Predictable Orbits

It's astonishing how quickly we learned to operate in space.

Sputnik I was launched in 1957. A half-dozen years later, there were tens of communication and navigation satellites being maintained in space. Within a dozen years, the number of space objects had exceeded one hundred and the U.S. had landed a man on the moon and returned him safely to earth. There now are hundreds of working spacecraft and thousands of pieces of space junk clustered mainly in three types of earth orbits.

The U.S. Navy began in 1964 to maintain a constellation of a half-dozen Transit satellites in 1000-kilometer high polar orbits. That meant establishing an infrastructure for routinely tracking each satellite, predicting its orbit, and uploading the data to the satellites, which broadcast them to the users. Positioning requirements of Transit were modest and Doppler positioning is forgiving of errors in satellite position. Good thing, too, because the satellite position could be off by kilometers in the early days.

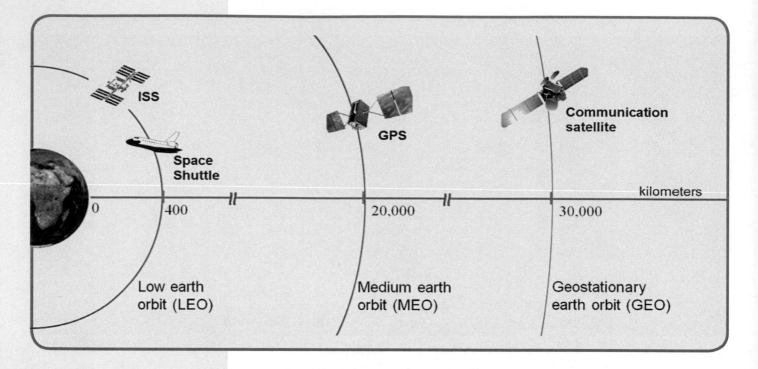

In the figure: ISS, Space Shuttle, GPS, Communication satellite, kilometers, 0, 400, 20,000, 30,000, Low earth orbit (LEO), Medium earth orbit (MEO), Geostationary earth orbit (GEO)

Equations of Motion

The satellites in space move in accordance with Newton's laws just as a ball tossed in the air does on the earth.

Recall Newton's Second Law: $F = m\,a$, where F is the applied force on an object of mass m, and a is the resultant acceleration. In order to predict the position of a satellite accurately, we just have to model the forces on a satellite accurately.

By orbit determination we mean a method to determine a satellite's position and velocity (the dynamical 'state') in a proper global coordinate frame and compiling a table, called an *ephemeris*, of time versus the state vector. We have already recounted the story of how Sputnik's orbit was determined from Doppler measurements. The early efforts in the 1960s in the U.S. to track its own satellites, and those of the USSR, led to development of both passive methods (e.g., imaging by camera, for angular measurements) and active methods (e.g., radar tracking and, later, laser ranging).

For real-time positioning, what's needed is to determine the state of each satellite at some epoch on the basis of measurements collected from tracking sites and extrapolate the orbit into the future. Determining the orbit of a satellite by fitting measurements to a model is not difficult, in principle. The challenge in the 1960s was to propagate the orbit to predict the future positions of the satellite. The theory was well known: Newton's laws of motion. The hard part was to understand the physical environment of the satellite and characterize the forces accurately.

The biggest force on a satellite is due to earth's gravitational pull. The gravitational pulls from the sun and moon are a millionth of that of the earth's, but must be accounted for. Even the pressure from photons from the sun impinging on the solar panels has to be taken into account if you want to predict a satellite's position 24-hours ahead with meter-level accuracy.

But the first challenge was to get the earth's gravity field right.

Shape of the Earth

If the earth were spherical in shape and uniform in composition, like a glass marble, its gravity field would have been simple. We could then model the earth as though its entire mass were concentrated at one point, its center, and the gravitational force exerted by it on a satellite would have been proportional to inverse-squared-distance measured from the point mass. If there were no other forces at work, a satellite orbit would have been fixed in space forever and predicting the satellites position at any instant *perfectly* would have been a snap.

Actually, the earth is neither spherical in shape nor uniform in composition and, as we have said, there are other perturbing forces on a satellite besides earth's gravity.

The earth is not spherical. It bulges somewhat at the equator and is flattened at the poles. Newton knew that. Measurements made just a few years after his death proved him right. The equatorial radius and polar radius differ by about 20 kilometers. That's not a lot when you consider that the equatorial radius is about 6,400 kilometers. The earth's shape resembles an ellipsoid generated by revolving an ellipse about its semi-minor axis. Of course, an ellipsoid is a smooth surface, and much of the earth isn't.

If the earth were indeed an ellipsoid, we'd see it in the surface of the oceans. Turns out the ocean surface is going up and down by up to 100 meters in relation to the smooth ellipsoid. We'll spend the rest of this section explaining that statement.

If the earth were an ellipsoid of uniform composition, we could take advantage of the symmetry to describe how its gravity field would change with distance from its center and the latitude, and capture it in an expression with a half-dozen terms. (Recall that the expression for the gravity field of a point mass changes with distance and has a single term.) No such luck!

The earth is an unsymmetric, nonuniform mess and three generations of post-Sputnik geodesists have devoted professional lifetimes to get a handle on its gravity. They have now got it nailed with sufficient precision so as to eliminate gravity uncertainty as a significant source of GPS positioning error. The expression for the earth's gravity field at a point contains thousands of terms.

Shape of the Earth

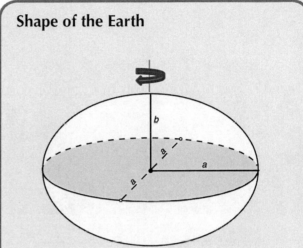

To get an idea of the shape of the earth, cut out from cardboard a scaled model of an ellipse with semi-major axis a – 6,400 kilometers and semi-minor axis b 20 kilometers shorter, glue a toothpick along the semi-minor axis, and spin the toothpick.

The shape you generate is the shape of the earth. It's called an oblate ellipsoid. (Spinning the ellipse about the semi-major axis would have given the more familiar egg shape, called prolate ellipsoid.)

The figure above greatly exaggerates the equatorial bulge.

Earth's Gravity Model

Characterization of the earth's gravity field required measuring its effect on satellite orbits and working backwards to determine what gravity forces must have been at work. But it was a classic Catch-22 situation: Accurate characterization of an orbit required accurate position estimates of the terrestrial tracking stations, which, in turn, required accurate satellite orbits. In the 1960s, NASA launched a program of geodetic satellites whose main mission was to focus on this bootstrapping process.

To get an idea of how complicated earth's gravity field is in space, let's look much closer—to the surface of the earth. It's easier to start with the ocean surface. If we disregard the waves and tides and define something called mean sea level, it would be a level surface: the gravity force at every point would be perpendicular to the surface. It's a surface of constant gravity potential, a subtle concept. Extended over land areas, this surface is called the *geoid*.

What does the geoid look like? Over the oceans the geoid coincides approximately with the mean sea surface. Over the continents it can be regarded as a continuation of the mean sea surface below the topography. Unlike the ellipsoid, it's an irregular surface. How close is the geoid to the ellipsoid? The geoid can be as low as 105 meters below the ellipsoid (at the southern tip of India) and as high as 85 meters above (around Papua New Guinea).

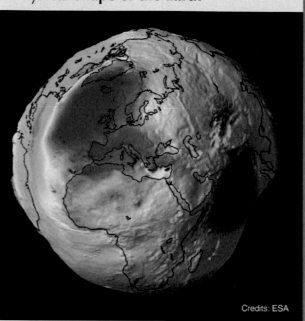

Physical Shape of the Earth

Credits: ESA

Physical shape of the earth is called the geoid, which extends the mean sea level surface over land areas.

The geoid relief in this figure has been artificially stretched for clarity of representation. In reality it ranges from -105 meters (shown in blue), to +85 meters (in red), which appear tiny when compared to the earth's radius ranging from 6,357 kilometers at the poles to 6,378 kilometers at the equator.

6.2 Global Coordinate Frame

We defined trilateration early on in this book as a method for determining your position by measuring distances to three objects whose positions were known to you. Later, we settled on a design of a global positioning system in which satellites were the objects at known positions, but we avoided the subject of how these positions—yours and the satellites'—are to be represented. We now tackle this topic.

While the position of each point is unique, its representation depends upon the coordinate reference frame in which it is expressed. Coordinate systems are said to be the *languages of location*. Many such languages and regional dialects came into being over the years and were adopted by different countries and geographic regions to meet their local needs. The need for a global coordinate frame arose with the coming of the space age (for representation of satellite orbits) and the requirements of globalization (for air travel and maritime commerce).

Position on a Sphere

The ancient Greeks knew the earth to be spherical. They had a good idea of its size (you can *google* Eratosthenes), and understood the concept of representing the position of a point on the earth's surface in terms of latitude, so many degrees north or south of the equator, and longitude, so many degrees east or west of some chosen meridian. You know about the equator; a meridian is a half-circle on the surface of the earth that goes through both poles. Points on the surface of the earth with the same latitude form a parallel; all points on a meridian have the same longitude.

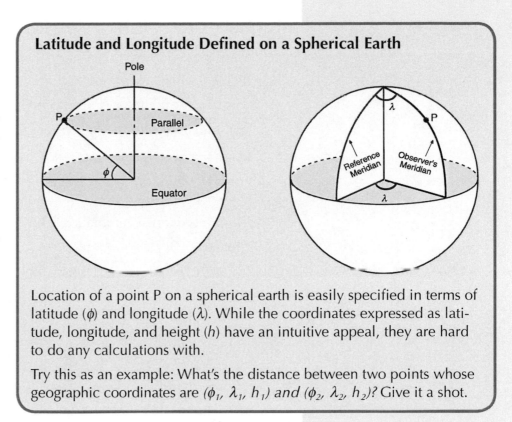

Latitude and Longitude Defined on a Spherical Earth

Location of a point P on a spherical earth is easily specified in terms of latitude (ϕ) and longitude (λ). While the coordinates expressed as latitude, longitude, and height (h) have an intuitive appeal, they are hard to do any calculations with.

Try this as an example: What's the distance between two points whose geographic coordinates are (ϕ_1, λ_1, h_1) and (ϕ_2, λ_2, h_2)? Give it a shot.

Throw in the height for a point above or below the earth's surface, and you are all set. Specify the latitude, longitude, and height of a point, and you've got it nailed. This so-called geographic coordinate system, with some modifications, is still in use. As we have noted, an ellipsoid provides a closer match to the shape of the earth than a sphere. This introduces some complications, but nothing major. The geographic coordinates are intuitive, but not easy to do calculations with.

As you'd recall from high school analytical geometry, you can set up a two-dimensional Cartesian coordinate frame on a page by picking an origin and drawing two lines through it at 90 degrees to each other, to be called x- and y-axes. You could express the position of any point on the page in terms of its (x, y) coordinates. A global positioning system requires us to extend this basic idea to three dimensions and global scale.

We need an origin, and definitions of x-, y-, and z-axes. We'd want the coordinates of points fixed to the earth to remain fixed and, therefore, we'd want the coordinate frame to be rigidly attached to the earth and rotate with it.

Earth-Centered, Earth-Fixed Cartesian Coordinate System

Let's define it simply as follows:

- The origin, we'll put it at the center of mass of the earth,

- z-axis coincides with the spin axis of the earth,

- x-axis coincides with the intersection of the meridian going through, say, the Greenwich Observatory, and the equator.

satellite coordinates: (X, Y, Z)

The y-axis gets defined automatically by convention of a right-handed system. We got ourselves a proper Cartesian coordinate system.

$$range = \sqrt{(X-x)^2 + (Y-y)^2 + (Z-z)^2}$$

It'd have worked, too, if we weren't so picky about meter-level accuracy. What's the problem? Well, for one thing, the spin axis is not fixed in relation to the solid earth. The pole of rotation, that's the point at which the spin axis can be thought to exit from the earth, can move several meters over a year. The phenomenon is called *polar motion*. That wouldn't have been a big deal for Columbus, but as the z-axis wanders, so would the equator, and the x-axis would wander, and our coordinate frame would become rickety and coordinates of points on earth could change over a year by meters. We can't allow that.

A solution is to nail the z-axis to the earth. We have historical records of polar motion and we'll define the z-axis as going through the mean position of the pole of rotation between years 1900 and 1905. We could have picked instead the pole of rotation for a particular epoch meaningful to you, or the mean pole between two other years, but let's not quibble over it. It's done. The main thing is that we have now fixed the z- and x-axes, and the y-axis is determined automatically by convention.

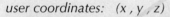

user coordinates: (x, y, z)

Realization of a Coordinate Frame

We defined a coordinate system, but it's all abstract. Who has seen the center of mass of the earth? And how are we to define the position of a point in this coordinate system? The answer to the second question is that we *realize* this abstract coordinate system by adopting the coordinates of a set of points and then refining them for *consistency* using astronomical methods, mainly satellite laser ranging and radar astronomy.

To see this idea at work, try a simple exercise. Take a clean sheet of paper, mark two points on it, and assign them coordinates arbitrarily. You have defined a *unique* coordinate frame: The location of the origin and orientations of x- and y-axes are uniquely determined. Can you pick coordinates of three or more points in this way and still obtain a unique coordinate frame?

The answer is yes, if the coordinates of points selected are *consistent*, i.e., each pair of points determines an identical coordinate frame.

In practice, the coordinates of the points selected would be determined from measurements (of ranges, angles, Doppler frequencies, etc.) and would have some error. We could still implement a coordinate frame as one that fits the data best in some sense, e.g., a least-squares fit. The more consistent the adopted coordinates, the more accurate the realization of the coordinate frame. First come coordinates of a set of points, then the coordinate frame. It's a bootstrapping process of definition and refinement for consistency.

The Cartesian coordinates, though easier for calculations, have problems of their own. So, your (x, y, z) coordinates are (in meters): (1 510 885, -4 463 460, 4 283 907). I can tell that you are in the northern hemisphere ($z > 0$), but beyond that I have no idea of your location. If I were handy with sines and cosines and could do quick calculations in my head, I'd say you are at mid-latitude, but can't tell if you are on the earth or in the air.

We started this chapter talking about latitude, longitude, and height, defined in relation to a spherical earth, and then complaining that calculations with these geographic coordinates were difficult. So, we defined a Cartesian coordinate frame, but that, too, didn't meet our expectation of ease of use. Turns out we need both coordinate frames, one for its intuitive appeal in everyday use and the other for calculations.

> **These are My GPS Coordinates**
> You mean these are my position coordinates in WGS 84 as provided by GPS.

WGS 84

Returning to our metaphor of a coordinate frame as a language of location, the *World Geodetic System* 1984, a global coordinate frame whose development by the U.S. DoD began in the 1950s, is the native language of GPS. With widespread use of GPS, WGS 84 now plays the same role that English does in global scientific communications. But the other languages have not disappeared, even within the U.S. The official coordinate frame of the U.S., adopted by a resolution of the Congress, remains North American Datum 1983, developed and maintained by the U.S. Department of Commerce.

The DoD, with its worldwide operations, needed a global coordinate frame in which to express the positions of its assets. This work started in the 1950s in Defense Mapping Agency and continues through its current incarnation as National Geospatial-Intelligence Agency. When somebody tells you "these are my GPS coordinates," what your interlocutor really means is: These are my position coordinates expressed in WGS 84, as provided by GPS.

Defining a global coordinate frame is a large undertaking that's never quite finished. You are always refining the coordinate frame to make it more consistent, and updating it to reflect changes wrought by the incessant creeping movement of tectonic plates

Straddling the famous brass strip at the Royal Observatory Greenwich, which is no longer the Prime Meridian of the World. (Wikimedia Commons photo)

1776 – John Harrison's H4

1928 -- Quartz crystal oscillator

1980 –The first GPS Rubidium atomic clock

adding up to several centimeters over a year, and larger, catastrophic displacements due to earthquakes. With the coming of GPS, the realization of WGS 84 has meant refinement of the coordinates of the GPS monitoring stations operated by the Control Segment.

Note that the prime meridian (or, zero meridian) is no longer defined by the famous brass line in the courtyard of the storied observatory in Greenwich, but on the basis of statistical calculations to reconcile the coordinates assigned to many points around the world. The zero meridian has moved from the brass strip eastward by about 102 meters, but we'd call it *Mean Greenwich Meridian* so as not to hurt the feelings of the Brits.

6.3 Perfect Clocks (Well, Almost)

Just a few hundred years ago, we would have been content to measure time well enough so as not to miss a planting season, and we now are talking in terms of billionths of a second to get through a routine day.

The clocks are the most important payload on a GPS satellite. They are not much to look at—well shielded rectangular boxes. They wouldn't remind you of your great grandfather's elegant pocket watch that fitted in his waist coat, or the grandfather clock in his living room. Chances are that your parents didn't own either a pocket watch or grandfather clock, and you look in your pocket for your cellphone when you need time, rather than bend your elbow to look at your wrist.

It's interesting that time, this quantity that troubled philosophers no end ("does time exist?"), is the one we can now measure most precisely.

About Clocks and Navigation

The intimate relationship between navigation and timekeeping has been known for centuries.

The difference in local times at two places is tied to the difference in their longitudes. Accurate clocks were known to be the key to finding longitude. Finding and keeping longitude to within one-half degree over a six-week voyage, as required by the Longitude Act (1714) of the British Parliament, meant that the cumulative timekeeping error could not exceed 2 minutes. The corresponding navigation error would have been 30 nautical miles (56 kilometers). In round numbers, the error budget was 3 seconds per day, an extraordinary challenge in the eighteenth century for a clock aboard a ship having to deal with changes in temperature and humidity, and rough seas.

GPS uses clocks differently.

GPS uses clocks to measure the transit time of a radio signal from a satellite to a receiver and obtains range by multiplying it by the speed of signal transmission. Radio waves travel in space at the speed of light and, in order to measure ranges with meter-level accuracy, transit times must be measured with an accuracy of a billionth of a second, which we'll now call a *nanosecond* (10^{-9} seconds). Such accuracy and precision are required in specifying both the time of transmission from a satellite and the time of arrival at the receiver.

A basic concept in the design of GPS is to keep the clocks in the satellites synchronized. The transmission time could then be marked on the signal in accordance with such clocks. Any drift in satellite clocks could be monitored from the ground and corrections uploaded to satellites periodically. For once-daily data uploads, the challenge faced by GPS in the 1970s was to develop space-qualified clocks that kept time to within several nanoseconds over a day.

About Clocks and Timekeeping

All clocks have two basic components: a pendulum or some type of oscillator to produce a regular set of 'ticks', and a way to count and display the number of these ticks. For an ideal clock, the ticking rate (or, frequency, f) of the oscillator and, equivalently, the interval between ticks (or, period, $1/f$), would be constant. The lower is the uncertainty in frequency, the higher the *accuracy*. The higher the frequency, the more finely a clock slices a second, and the more *precise* is the time measurement.

The requirements of radio communications early in the 20th century led to development of quartz crystal oscillators with fractional frequency uncertainty of about one part in a million, which kept time on average to within 0.1 second over a day.

Chances are your wrist watch has a quartz oscillator with a resonant frequency of 32,768 hertz (2^{15} hertz) and a digital circuit generates a one-second pulse by counting 2^{15} oscillations. The precision internal to the watch is 2^{-15} seconds, about thirty-millionths of a second, not precise enough for measuring GPS signal transit time.

For nanosecond-level precision and 10-nanosecond accuracy over a day, we need an oscillator that ticks at least a billion times a second and has frequency stability of one part in 10^{13}. There is no hope of meeting such requirements with electro-mechanical oscillators, which are influenced by atmospheric conditions, aging, and wear.

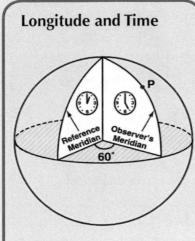

Longitude and Time

The difference in local times at two places is tied to the difference in their longitudes

How much frequency uncertainty can we tolerate in a satellite clock?

Suppose we decide for practical reasons to upload data to GPS satellites no more than once a day.

For meter-level positioning, roughly speaking, we don't want the error (Δt) to build up to more than 10 nanoseconds over a day ($t = 10^5$ seconds, in round numbers).

So, how much frequency uncertainty can we tolerate in the oscillator with advertised frequency of f?

Let's look at the fractional error

$$\frac{\Delta f}{f} = \frac{\Delta t}{t} = \frac{10\,\text{nanoseconds}}{1\,\text{day}}$$

$$\approx \frac{10 \times 10^{-9}\,\text{seconds}}{10^5\,\text{seconds}} = 10^{-13}$$

The answer is: one part in 10^{13}.

I.I. Rabi

Invention of Atomic Clocks

The revolution in timekeeping began in mid-1940s when Isidor Isaac Rabi, a physics professor at Columbia University, suggested that quantum-mechanical behavior of atoms could provide an oscillator that was largely immune to external influences.

Quantum mechanics confines electrons bound to an atom to certain specific energy states, and exciting the atoms with the 'right' electromagnetic radiation can cause the electrons to jump from one state to another. In cesium (atomic number 55; symbol: Cs) there is one such jump, called a hyperfine transition, which occurs when an atom is hit with microwaves with frequency 9,192,631,770 hertz. So, to use cesium as an atomic clock, what is required is to scan cesium atoms with a range of frequencies until such transition is 'seen' to occur, and then to hold the microwave emitter at that frequency using principles of feedback control.

The microwave generator in this scheme, operating perfectly in sync with the resonant frequency of the atoms, is the atomic clock. In round numbers, it ticks at a rate of 10 gigahertz, has a period of 0.1 nanoseconds.

Rabi won the 1944 Nobel Prize for Physics.

GPS, the Time Giver

Atomic clocks have gotten a lot better over the past 25 years, but this is not a book about clocks and timekeeping and we wouldn't obsess over a billionth versus a trillionth of a second. We should, however, point out that atomic clocks keep time by exploiting properties of the outer electrons of an atom. There is nothing dangerous going on. Playing with the nucleus could be dangerous, but we stay away from it.

The rubidium and cesium atomic clocks available commercially in the 1970s offered frequency stability of about one part in 10^{13}, and were capable of keeping time to within 10 nanoseconds over a day. While such stability met the requirements of GPS, the challenge lay in re-packaging them for radiation hardening to operate in the space environment. GPS orbits transit the upper Van Allen Belt of high-energy charged particles. All electronic equipment requires radiation hardening to prevent damage.

Your one-dollar GPS receiver gives you access to the clocks aboard the satellites. You don't need to spend thousands of dollars for an atomic clock of your own. The irony is that such easy access to precise time has wiped out the market for atomic clocks to the extent that the GPS program now has trouble procuring the necessary clocks for its satellites.

Physics of Atomic Clocks

According to the laws of quantum physics, atoms absorb or emit electromagnetic energy in discrete amounts (or quanta) that correspond to the differences in the energy between the different configurations of the electrons surrounding the nucleus.

An atom can make a transition between two energy states E_0 and E_1 by absorption or emission of energy in the form of electromagnetic radiation having the precise frequency

$$v_0 = |E_1 - E_0| / h$$

where h is the Planck constant and v_0 is called the atomic resonance frequency.

Cesium 133 atoms have a resonance frequency of exactly 9,192,631,770 hertz. The period of this wave is close to a tenth of a billionth of a second.

The cesium fountain standard NIST–F1 can keep time to within 1 second in 100 million years (if run continuously).

(Geoffrey Wheeler Photography)

We talked about GPS first in terms of broad principles and then in terms of technologies that made its realization possible. We now are ready to describe the system for real.

The GPS infrastructure basically has two parts that the U.S. Government manages: The part in the sky, called the Space Segment, that transmits radio signals at the users, and the part on the ground, called the Control Segment, which keeps an eye on the part in the sky and prompts it, as necessary. We'll describe these briefly below.

The third part covers the all-important GPS receivers and is called the User Segment in program managers' lingo, but the U.S. Government has no role in the civil receiver market. We'll talk below a bit about the evolution of GPS receivers over the past 30 years.

Let's pause briefly to survey the state of the technology that would have been available to the developers of GPS in the mid-1970s. An easy way is simply to point out things that didn't yet exist: personal computers were 10 years away; Internet was 15 years away; cellphones were 20 years away; and smartphones were 40 years away.

Let's now talk about the architecture of GPS.

Space Segment: Satellites

The Space Segment comprises a constellation of about two dozen satellites. The actual number has remained between 25 and 30 since 1995. The satellites are arranged in nearly circular orbits in six planes. Each orbital plane is inclined at 55 degrees relative to the equatorial plane. The altitude of the satellites is about 20,000 kilometers, or about three earth radii.

Each satellite circles the earth twice each day, moving at the rate of about 4 kilometers per second. That's about 15 times faster than an airliner at cruising speed. The orbital period of a satellite is 11 hours 58 minutes. That's how long it takes the earth to turn 180 degrees on its axis and, at the end of two orbits, a satellite repeats its ground track.

A constellation of 24-plus satellites allows most users with an unobstructed view of the sky to 'see' four or more satellites. As we'll see, four is the magic

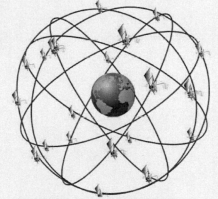

The baseline GPS satellite constellation consists of 24 satellites in inclined orbits 20,000-kilometers high.
(FAA graphic)

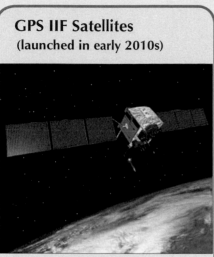
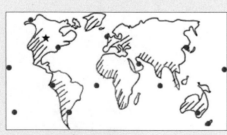

GPS signals are monitored continuously from stations located around the world (www.gograph.com)

number. If the number of satellites in when you pull off an open highway int

A GPS satellite with solar panels folded in is about the size and weight of a Ford Focus sedan. The satellites being launched in the early 2010s are from a batch of a dozen designed and built by Boeing and called Block IIFs. These constitute the fourth generation of GPS satellites. The earlier batches bore prosaic names Block I, Block II, and Block IIR, as you would have guessed if you know any engineers.

The GPS satellites have performed like the Energizer bunny—they just go and go, years past their design lives. That's a mixed blessing: GPS managers would like to replace the early models with newer, more capable satellites, but can't make a case to the Congress for replacing working satellites.

GPS Block IIR satellite (Lockheed Martin photo)

Starting with the first experimental satellite in 1978, a total of about 60 GPS satellites have been launched, of which 30 remain in active service in the summer of 2015.

Control Segment: Day-to-Day System Operation

The Control Segment performs the unglamorous but vital function of operating the system from day to day. That means constantly watching the system to ensure that it is operating within specifications, periodically uploading new data on orbits and accumulated clock biases for the satellites to transmit, and occasionally moving satellites to maintain them in their assigned slots in space.

The Control Segment comprises (i) about a dozen unmanned sites scattered around the world from which to monitor the satellite signals using receivers like the one you have, but more precise in their measurements, (ii) a Master Control Station at Schriever Air Force Base outside Colorado Springs to analyze these measurements for system performance, and (iii) several antennas at different locations from which to upload commands and data to the satellites.

It surprises people to see antique computers and other electronics at government facilities dealing with advanced technologies. The Master Control Station used until 2009 IBM mainframe computers from the 1970s.

GPS is controlled in near real time from Schriever AFB outside Colorado Springs. If you were allowed in the GPS Master Control Station, you'd see seven young Airmen of the 2nd Space Operations Squadron (2SOPS), average age 23, 'flying' GPS satellites.

There would be three space system operators, one space vehicle operator, one payload system operator, one network administrative operator, and one mission chief.

It's a lean operation.

(U.S. Air Force photo)

GPS Master Control Station

The Control Segment draws attention only when things go wrong. It is to the credit of the U.S. Air Force Space Command that GPS has performed nearly flawlessly. There have been a few screw ups with satellite clocks going bad and wrong data being uploaded. If an irregularity is detected by a monitoring station, the first corrective action is to quickly upload a message to the satellite to change the health bit in its broadcast to 'unhealthy' so a receiver would know to disregard the signal. In time, this would be accomplished in seconds; up to now, it has taken minutes, even hours.

An out-of-spec signal from a satellite while your airplane is making a GPS-guided approach to a fog-bound SFO can cause havoc, but you don't have to worry about it. Civil aviation requires that GPS receivers validate each position estimate by cross-checking the range measurements for consistency. A typical consumer receiver wouldn't bother with such computation-heavy integrity monitoring of the signals.

Thankfully, the rate of anomalous events has been going down and is now about one a year.

New, Improved GPS

GPS is being modernized.

The process began in 1998. Well, modernization may not be quite the right word. You don't start modernizing a system that became operational barely three years earlier and is performing like gangbusters. But engineers aren't good with words and politicians need slogans, so let's not quibble over it.

When it became clear that GPS was going to be a lot more useful than originally thought, it was decided to add more features to it. That basically meant more signals designed to be more robust. Of course, that

GPS III satellite now under development (Lockheed-Martin graphic)

would require a re-design of the satellites and the Control Segment, but it would all be worth it.

It would be 2020 by the time all elements of the modernized GPS are in place.

User Segment: GPS Receivers

GPS military receivers circa 1982 (U.S. Air Force photo)

The Space Segment and Ground Segment have remained in form and function substantially as designed 40 years ago, but the User Segment has seen a revolution, or two. The receivers of today bear no resemblance to the early receivers.

The first mobile, commercial GPS receivers appeared in the early 1980s: the Texas Instruments TI 4100, developed under the leadership of Phillip W. Ward; and the Macrometer V-1000, designed by Charles C. Counselman III, an MIT professor, and commercialized by Litton Aero Services. These receivers cost about $150,000, and were intended for surveyors and geophysicists who could use the GPS signals then available from a partial constellation of Block I satellites to determine relative positions of two points very precisely—at centimeter level. The answers were not needed in real time and the surveyors could schedule data collection at a site when four or more satellites would be in view. With a 10-satellite constellation in the early 1980s, that invariably happened between 2 a.m. and 4 a.m., regardless of place, as old timers now recall.

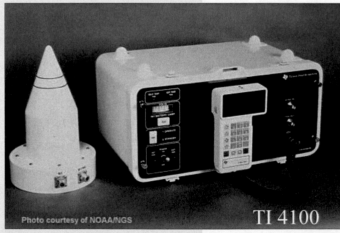

Photo courtesy of NOAA/NGS

TI 4100

The TI 4100 and Macrometer were revolutionary instruments. Laborious surveying work that previously took a week and required a team could be accomplished in half a day by one person. The theodolites and chains used by surveyors for a century would soon become relics. More compact GPS receivers for surveyors came on the market in the late 1980s with prices only one-tenth of those of a TI 4100 and Macrometer.

In the early 1980s, if you wanted a GPS receiver, you could buy one from Texas Instruments -- for $150,000.

There is a story told about a Dallas oilman who heard about these TI boxes that could keep him from getting lost. He called up the CEO of TI to ask about having one built for his Cadillac. The CEO is reported to have responded that if the oilman bought one of his GPS receivers, TI would throw in a Cadillac for free. (Actually, the oilman would still have gotten lost because digital street maps didn't yet exist.)

The Chip and Digital Maps

Microelectronics changed the touch and feel of all electronic devices in the late 1980s, GPS receivers included. Chances are you wouldn't even have heard of GPS if it weren't for the chip.

The first commercial handheld unit appeared in 1988: the Magellan NAV 1000. It weighed about a kilogram, cost a couple of thousand dollars, and was aimed at hikers and boaters. The tiny screen could only show text—latitude and longitude. There was no room or resolution to display a map. It didn't matter because digital maps didn't yet exist. But the text display is all you needed to navigate in a featureless desert and thousands of NAV 1000s were bought and shipped to soldiers, often by their families, during the Gulf War in 1991. An Army Jeep delivering mail or meals could drive straight to the right tent with little fuss. GPS was an instant hit with the troops.

The making of digital maps took off in the mid-1980s and shows no sign of abating. First there were street maps intended for drivers of cars and trucks and charts for the boaters, all based on static data bases. Now we have dynamic data bases, but more on them later.

Magellan NAV 1000
(Magellan photo)

Motorola demonstrated a GPS multi-chip module receiver in 1993, followed by several companies that started producing GPS receivers built with a small number of integrated circuits. It became common to integrate RF and IF functions on a single application-specific integrated circuit generating a stream of digital output. Then came a spate of so-called personal navigation devices with large, high-resolution screens displaying your position on a roadmap.

A particularly successful maker of such devices was Garmin, a company founded by Gary Burrell, a marketing guy, and Min Kao, a Ph.D. electrical engineer. The name Garmin is a contraction of their first names. There was competition from TomTom, Magellan, and others. Some 6 million units to be propped up on dashboards were sold in 2007, and 8 million in 2008. But the heyday of portable, stand-alone GPS units was already over.

Smartphones Take Over

A GPS receiver is a one-way radio device, but that's not enough in an increasingly connected world. The new generation of devices must connect to the Internet and to each other. Specialized products like Dick Tracy-style GPS watches for runners or kayakers will continue to appear, but smartphones with displays the size of old Garmins and TomToms are the ideal platform for a GPS receiver on a chip.

It's not enough to know there is a gas station five miles down the highway; you want to know if it's open, considering it's 11 p.m. on a Sunday. And there is a hotel nearby, but you want to know whether it has a vacancy, and you can afford their rate, and what did previous customers think of their service. A smartphone can tell you everything.

A new thing is dynamic data bases built around *current* contributions from cellphone users on the scene. Google recently announced acquisition of Waze Mobile, a 5-year-old Israeli start-up, which has developed real-time mapping of road traffic with contributions from drivers. Reported price tag: $1 billion. (The same week, it was announced that Jeff Bezos of amazon.com fame had bought an American institution called *The Washington Post* for $250 million in cash. You have to be of a certain age to appreciate the irony.)

GPS Signals

We have been slippery when it came to talking about GPS signals.

And for a good reason: The radio signals are hard to describe in everyday language. We bought time by simply comparing them to signals broadcast by the commercial AM/FM stations, but we now have to bite the bullet and talk seriously about this central magic of GPS.

The AM/FM analogy is imperfect. Each AM/FM radio station is assigned a distinct frequency and the signals have a simple structure: an audio signal 'riding' on top of a radio wave of assigned frequency called a carrier. The radio signal changes continuously and each receiver extracts from it an *identical* audio signal and feeds it into a speaker.

The radio signal transmitted by a GPS satellite is repetitive and monotonous. All satellites transmit at the same frequency, but the signal transmitted by each has a unique structure, which is known to the receiver. A GPS receiver determines from each signal a *unique* measure of its range to the satellite and combines such measurements from four or more satellites to compute its *unique* position. It will take us three chapters to tell the complete story.

We'll start with the signal structure.

While the number of signals transmitted by a GPS satellite has grown over the past 10 years, we limit ourselves to the lone signal that has been the mainstay for civil applications from the beginning. We'll call it the *open* signal. When we say GPS signals, we mean the open signals transmitted by all satellites.

Two main things we want to emphasize in this chapter are: first, the GPS signals are extremely weak; and second, what the signals lack in raw power, they make up in ingenuity of design.

GPS Signals are Extremely Weak

That's not a surprise.

We have a 30-watt transmitter aboard a GPS satellite 20,000 kilometers away. How much is 30 watts? That's the power you burn in a car headlight and, like a headlight, the transmitting antenna in the satellite focuses the power in a beam. A GPS satellite illuminates about a third of the earth's surface below, spreading its 30 watts more or less evenly.

> ### GPS Open Signal
> The first GPS satellites transmitted three signals split between two frequencies. There was an open signal for civil users and two encrypted signals for users authorized by the DoD.
>
> The satellites being launched in 2015 transmit seven signals split among three frequencies.
>
> We focus in this book exclusively on the open, legacy signal, which found a thousand uses and a billion users.

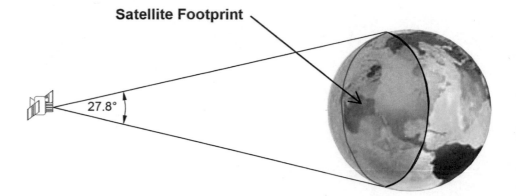

Satellite Footprint

27.8°

A GPS satellite illuminates about one-third of the earth's surface (DoD graphic)

The signal power reaching your receiver is only about 10^{-16} watts, as in a tenth of a millionth of a billionth of a watt (0.000 000 000 000 000 1 watts). While that may seem very small, it's not immediately obvious that we have a problem.

After all we get TV signals beamed from geostationary satellites that are even farther away. Yes, these TV signals are comparable in strength, but they require a large dish antenna, impractical for a pocket GPS receiver or cellphone. On the other hand, while TV signals are unpredictable, the structure of a GPS signal is essentially known, except for its phase at the antenna, which is tied to its transit time. If you understood that statement, you can move on to the next chapter.

GPS signals are designed to overcome the power deficit in outdoor use with clear lines of sight to satellites. If a signal is obstructed, it would be weakened further. That would happen, for example, when you are hiking in the woods with a dense vegetation canopy. You may have to step off the trail into a clearing to determine your position. A metal or concrete structure can block the signals nearly completely. GPS is essentially unusable in tunnels and underground parking garages.

What Do We Want from GPS Signals?

We have been dropping hints here and there, but let's now lay down the requirements straight.

- We require that all satellites transmit at the same frequency. We have to listen to signals from multiple satellites simultaneously and don't want to have to tune to each separately. So, the signals have to be such that we can listen to each satellite without interference from the others.

- We require that the signal arrival time be determined with an accuracy of nanoseconds (as in billionths of a second). It follows because we want to determine ranges to the satellites within meters.

- Now the clincher: The signals would be very weak, but there is no room to mount a big antenna on the receiver.

That seems like a tall order. We have to find some leverage to tackle the problem.

Large Spectral Real Estate to Spread Out on

The leverage comes from getting hold of a big chunk of the radio spectrum —24 megahertz wide. For comparison, note that AM stations in the U.S. get 10 kilohertz each, and FM stations get 200 kilohertz each. A GPS signal ties up more frequencies than the entire FM band.

The radio spectrum is a finite and precious resource, getting more precious as our appetite for personal mobile communications grows. The spectrum is managed at the international level by the International Telecommunication Union, a specialized agency of the U.N., and at the national level by the Federal Communications Commission. The demand keeps growing and the supply doesn't. We have to be smart about utilizing the radio spectrum.

The 1940s and '50s were horse-and-buggy times when it came to wireless communications. The demands were simple, in general, though the military has always been sensitive about its communications falling into the wrong hands. There was a lot of unused radio spectrum available and it was natural to think that we may be able buy message security by using up more of the spectrum than was actually needed. So, the U.S. military developed during WWII a new class of signals, called spread spectrum signals.

The idea was to spread the signal power over a much larger frequency band than

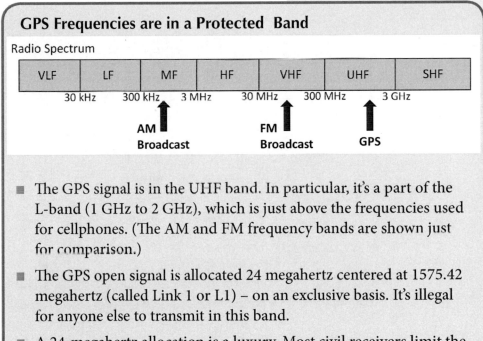

GPS Frequencies are in a Protected Band

Radio Spectrum

| VLF | LF | MF | HF | VHF | UHF | SHF |

30 kHz   300 kHz   3 MHz   30 MHz   300 MHz   3 GHz

AM Broadcast   FM Broadcast   GPS

- The GPS signal is in the UHF band. In particular, it's a part of the L-band (1 GHz to 2 GHz), which is just above the frequencies used for cellphones. (The AM and FM frequency bands are shown just for comparison.)

- The GPS open signal is allocated 24 megahertz centered at 1575.42 megahertz (called Link 1 or L1) – on an exclusive basis. It's illegal for anyone else to transmit in this band.

- A 24-megahertz allocation is a luxury. Most civil receivers limit the signal to 2 - 4 megahertz width.

was needed to transmit the signal. This spreading was done by introducing deliberate, rapid phase variations in the signal by multiplying it with a binary-valued (i.e., zero or one) code, called a *spreading code*. The faster the code (in bits per second), the more the signal spreads out in the frequency domain. The intended receiver would know the code and could undo this operation (i.e., *de-spread* the signal) to retrieve the original message. Message security was bought by keeping the codes secret.

By selecting the codes judiciously, many transmitter-receiver pairs could use the same radio frequencies, as we discuss below. After the War, spread spectrum signaling was adapted for radar, and later for satellite navigation.

A GPS satellite has to tell you about its health status, ephemeris (position and velocity), and clock bias parameters. This can be accomplished with a low data rate, say, 25 to 100 bits per second, requiring 50 hertz to 200 hertz of the radio spectrum. Let's now talk about spreading out this signal over many megahertz and the benefits of this operation.

Pseudo-Random Noise Codes

The spreading codes of interest to us belong to the extended family of pseudo-random noise (PRN) codes, so called because their binary sequences have an appearance as though generated by the outcomes of random events, like coin tosses. Here's an example:

... , 0, 0, 0, 1, 0, 0, 1, 0, 1, 1, 0, 0, 1, 1, 1, 1, 1, 0, 0, 0, 1, 1, 0, 1, 1, 1, 1, 0, ...

Actually, these sequences are generated by a carefully designed algorithm and have very special mathematical properties.

Each GPS satellite is assigned a unique code for its open signal from a sub-family called C/A-codes (for Coarse/Acquisition-codes; the name is unimportant for our purpose). Each C/A-code is of length 1023, and repeats each millisecond. The bit rate is 1.023 million bits per second, spreading the signal power mostly over a band of frequencies 2.046-megahertz wide. (The military signals utilize more of the 24-megahertz allocation.) As with any periodic signal, we can define the *phase* of a C/A-code by the bit number, going from zero to 1022, and back to zero.

We now define *correlation* as a measure of similarity between two PRN sequences. Write one below the other and count the number of agreements and the number of disagreements. Clearly, if the two sequences are identical, all bits would agree and we'd say they are perfectly correlated (correlation = 1). If the number of agreements is equal to the number of disagreements, we'd call the sequences uncorrelated (correlation = 0).

Ideally, we'd like a family of PRN codes to display the following properties: (i) in order to determine the arrival time of a signal accurately, we'd like each code to be uncorrelated with any delayed version of itself, and (ii) in order for the signals not to interfere with each other, we'd like each code to be uncorrelated with all others regardless of the alignment (see sidebar). We'll discuss in the next chapter how a receiver exploits these properties.

Correlations among Binary Codes

Length-1023 codes used for the GPS open signals, though considered short, are still too long for us to play with. So, we'll look instead at length-31 toy codes from a related family to analyze their correlation properties.

Here are two members of the family:

C1: 1, 0, 0, 0, 0, 1, 0, 0, 1, 0, 1, 1, 0, 0, 1, 1, 1, 1, 1, 0, 0, 0, 1, 1, 0, 1, 1, 1, 0, 1, 0

C2: 1, 0, 0, 0, 0, 1, 1, 0, 0, 1, 0, 0, 1, 1, 1, 1, 1, 0, 1, 1, 1, 0, 0, 0, 1, 0, 1, 0, 1, 1, 0

Recall that we define correlation as a measure of similarity between two binary sequences. By definition, C1 is perfectly correlated with itself. Correlation: 31/31

Let's now look at C1 and C1-delayed-by-one-bit, denoted as C1(+1):

C1:   1, 0, 0, 0, 0, 1, 0, 0, 1, 0, 1, 1, 0, 0, 1, 1, 1, 1, 1, 0, 0, 0, 1, 1, 0, 1, 1, 1, 0, 1, 0

C1(+1): 0, 1, 0, 0, 0, 0, 1, 0, 0, 1, 0, 1, 1, 0, 0, 1, 1, 1, 1, 1, 0, 0, 0, 1, 1, 0, 1, 1, 1, 0, 1

Correlation: (number of similar bits – number of dissimilar bits)/31 = - 1/31

Ideally, we would have wanted this correlation to be zero, but that's all we can get out of a length-31 PRN sequence.

Let's now look at correlation between C1 and C2.

C1: 1, 0, 0, 0, 0, 1, 0, 0, 1, 0, 1, 1, 0, 0, 1, 1, 1, 1, 1, 0, 0, 0, 1, 1, 0, 1, 1, 1, 0, 1, 0

C2: 1, 0, 0, 0, 0, 1, 1, 0, 0, 1, 0, 0, 1, 1, 1, 1, 1, 0, 1, 1, 1, 0, 0, 0, 1, 0, 1, 0, 1, 1, 0

Correlation: - 1/31

Suggested exercise: Create delayed or advanced sequences C1($\pm$ n) and check out their correlations with C1, and correlation between C1($\pm$ n) and C2($\pm$ m), for any values of n and m. All are uniformly low.

The longer the code, the closer its correlation properties are to the ideal. But there is a price: the longer the code, the higher the computational load on the receiver. The C/A-code family with its length-1023 codes represents a tradeoff, given the technology available in the 1970s.

Let's now pick up a detail we have avoided until now. Each satellite transmits a binary navigation message containing data to be used by the receiver for determining satellite health, position, accumulated clock bias, etc. The data rate is slow: 50 bits per second.

To summarize, a GPS signal has three components: carrier (sinusoidal wave of frequency 1575.42 megahertz); binary code (random-looking sequence of 0s and 1s with a rate of about 1 megabit per second); and navigation message (50 bits per second). Next we see how the three components are combined to create the signal.

GPS Signal Leaving the Satellite Antenna

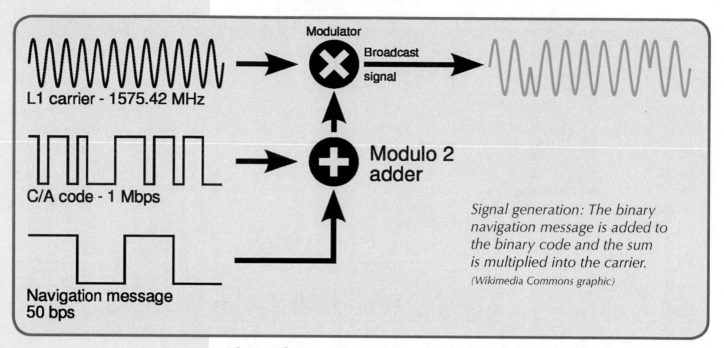

Modulator

Broadcast signal

L1 carrier - 1575.42 MHz

C/A code - 1 Mbps

Modulo 2 adder

Navigation message 50 bps

Signal generation: The binary navigation message is added to the binary code and the sum is multiplied into the carrier.
(Wikimedia Commons graphic)

The two binary components are added together (modulo 2) and the result is imprinted on the carrier. The technical term is *modulation*. As a communication engineer would say, we want to modulate the carrier wave with a binary-valued sequence.

We define a radio pulse, with positive voltage for one value and negative for the other. The pulse width would be determined by the code rate: about 1-microsecond wide for the approximately 1 megabit per second rate of the C/A-codes. The binary component is now represented as a sequence of rectangular pulses. Multiply this pulse train into the carrier wave and what you get is a sinusoidal function with 180-degree phase discontinuities corresponding to the edges of the pulses. That's the modulated signal.

We can now visualize a GPS signal radiated from a satellite antenna frozen in space for a moment. The signal looks like a sinusoidal wave with quasi-periodic phase reversals. The information carried by the signal is all in the locations of these phase flips.

What Goes on Inside a GPS Receiver?

While the look and feel of a GPS receiver has changed a great deal in the past 25 years.

But what goes on inside hasn't changed much. The basic steps of signal processing are the same, but are now implemented using smaller electronic components that are easy on power consumption.

GPS receiver circa 1985
(U.S. Air Force photo)

GPS Signal at the Receiver

Let's pick up the radio signal at the satellite antenna from the last chapter and follow it down to the receiver. With a power of 30 watts, it's a good, strong radio signal leaving the satellite. We have already talked about its structure: a radio wave of frequency 1575.42 megahertz with phase discontinuities characteristic of the binary pseudo-random noise code assigned to the satellite.

GPS receiver chips for cellphones circa 2010 (SkyTraq photo)

How does the received signal differ from the one that's transmitted?

We expect the received signal to be a delayed and weakened version of what was transmitted 70 to 90 milliseconds earlier, though the signal frequency would have changed —Doppler-shifted because the satellite is moving and the receiver may be on a mobile platform. But look at the signal reaching the receiver in your smartphone. It appears to have no discernible structure.

What happened to the GPS signal? Well, it's still there, but having spread out over a third of the earth's surface, it's now very weak. On top of it, it's overwhelmed by noise – the unwanted, unstructured radio signals seen by the antenna and thermal noise created by certain components of the receiver itself.

Radio noise is everywhere. You hear it if you tune an AM/FM radio to an unassigned frequency. On TV, the noise appears as snow. It is present at assigned frequencies as well, but the signal is so much stronger that the noise doesn't draw attention.

We spread the signal out over frequencies covering 2 megahertz – remember, spread spectrum signaling—and are now stuck with the noise present in that large band. The noise is 100 times stronger than the signal. The ratio of signal power to noise power is 0.01 or, as a communication engineer would say, the signal-to-noise ratio (SNR) is -20 dB. It's no wonder we can't see the GPS signal.

There is no hope of extracting the signal buried so deep in noise. But we don't have to extract the signal. We already know what it looks like. What we don't know is its phase at the antenna. The main task of the receiver is to leverage its knowledge of the signal structure (i.e., the pseudo-random noise code assigned to the satellite) to sense the phase of the received signal: which edge of the code is hitting the antenna?

Sensing the Phase of the Received Signal

While that seems like a challenge for a signal with an SNR of -20 dB, remember that we designed the signal specifically with this goal on mind.

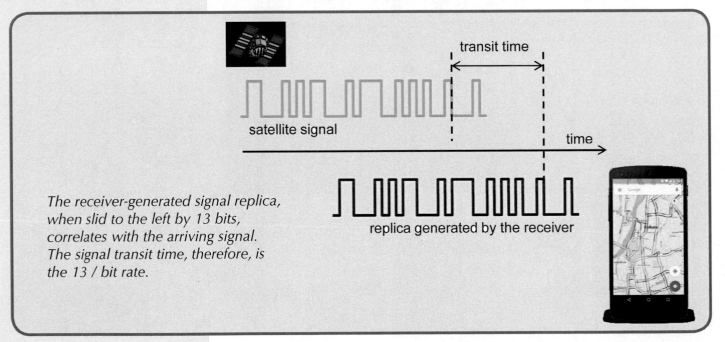

The receiver-generated signal replica, when slid to the left by 13 bits, correlates with the arriving signal. The signal transit time, therefore, is the 13 / bit rate.

Here's how it's done: the receiver generates a replica of the known binary code and attempts to align it with the code of the incoming signal by sliding it back and forth to obtain a match. To return to the vocabulary introduced in the previous chapter, the receiver correlates the arriving signal with the receiver-generated replica whose phase is varied systematically. This process consists of multiplying the incoming signal with the replica and integrating or summing the energy. The receivers have thousands of correlators that work in parallel to expedite this step.

If the signal and replica are misaligned by even a little bit, the energy in the individual pulses essentially cancels and correlation is near-zero. When they are perfectly aligned, the energy adds constructively and builds up to a peak that stands out. The phase of the corresponding replica essentially tells the receiver all it needs to determine the transit time of the signal.

Correlators are at the heart of a GPS receiver. But the incoming signal first has to be conditioned for processing before correlation can be attempted. We take that up next.

What We Want from a GPS Receiver

The answer is: a lot. Here's a list:

- Gather radio signals in a 2-megahertz band centered at 1575.42 megahertz and prepare them for processing.

- Determine which satellites are in view and sense the phase of each arriving signal. Determine the transmit time and receive time for each. (Transmit and receive times are entrenched terms – radio engineers are not bound by the rules of grammar.)

- Keep track of the changing phases of the signals and output transit times, say, once per second.

- Measure the Doppler frequency for range rates to the satellites.

- Decode the 50-bits-per-second navigation message to determine satellite position, velocity, and clock parameters.

- Calculate your position, velocity, and time.

We'll look at a functional description below of a generic receiver to accomplish these goals.

The GPS chip in your smartphone is about the size of the nail on your little finger. An advantage this chip has is that it can tap into other resources required by the phone anyway: clock, display, power supply. But there is no getting around the antenna. A GPS antenna is crammed in there, along with the antennas for GSM or WCDMA, Wi-Fi, Bluetooth, NFC, etc.

The job of a GPS receiver in a smartphone is made substantially easier by the cell tower, which has a GPS receiver of its own. A tower 10 kilometers away mostly sees the same satellites as your receiver and can download in a flash a data set of several hundred bytes, sparing your receiver the tedious and time-consuming chores like deciphering the molasses-slow satellite navigation message for ephemeris and other parameters. This mode of operation has a name: *assisted-GPS*.

Receiver Components by Function

All radio receivers have some things in common. All require an antenna to capture radio signals in a frequency band of interest. All have what are called front ends, which condition the radio signal for processing and typically include band-pass filters to separate the signal of interest from the others by frequency selection; amplifiers to jack up the signal power; and down-converters to bring down the carrier frequency from RF to something more manageable.

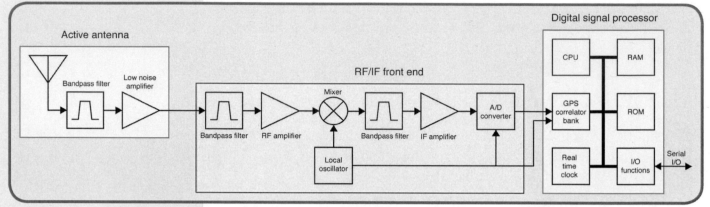

GPS receiver schematic
(Richard Langley, GPS World)

Band-pass filters in a GPS receiver trim the signal in frequency, keeping only about 2-4 megahertz-wide part around 1575.42 megahertz. The amplifiers jack up signal power a million times, or more. The noise is jacked up correspondingly. The signal-to-noise ratio is still 0.01 or -20 dB. Down-converters pull the center frequency down from 1575.42 megahertz to around 20 megahertz. The signal is still continuous, or analog, but not for long.

Signal processing is mostly digital now, given cheap microprocessors. So, next comes a digitizer or analog-to-digital converter. That means sampling and digitizing the continuous signals. The typical sampling rate is 4-8 megahertz. If you know of Shannon's sampling theorem, you know that you have to sample at least at the rate of twice the signal bandwidth, or 4 megahertz. An interesting part is that most GPS receivers for the consumer market digitize the signal at 1 bit per sample—a zero or a one. The only information you have in each sample is whether the signal was positive or negative. That's all the receiver is left to work with. That's all it needs.

The cell tower would already have told the receiver which satellites to look for. Next comes the heavy-duty computational work done by the correlators to acquire the signals and track them continuously. The final step is for a microprocessor to calculate position, velocity, and time. We'll save that for the next chapter.

To summarize, a receiver determines the transmit and receive times of each signal. Their difference gives the transit time. Multiply that by the speed of light and you get – well, it depends.

You Sure You Want to Measure Ranges to Satellites?

It seems trilateration requires it, but it will cost you. Let's take another look.

If you want to measure ranges to GPS satellites, your receiver clock would have to be synchronized with the atomic clocks in satellites which keep time nearly perfectly. But perfection doesn't come cheap. If a receiver required an atomic clock and cost $10 000, there wouldn't be a billion users, now would there?

Turns out you can get by with a cheap crystal oscillator costing just a couple of dollars. And, when you need time, you don't have to rely on your cheap clock – the receiver can output near-perfect time obtained directly from the satellite clocks. Of course, there is a price – there always is – but not up front and not in dollars.

Using an inexpensive receiver clock in the receiver, your transit time measurements from different satellites would have a common bias or offset, the size of which would depend upon the extent your receiver clock is running fast or slow at that instant in relation to the synchronized clocks in the satellites. Let's call them pseudo-transit times. Multiply them by the speed of light, and you get range measurements with a common bias, to be called *pseudo-ranges.*

We can work with pseudo-ranges for positioning. The price we have to pay is that we now have to contend with a fourth unknown: receiver clock bias at the instant the measurements were taken. That's in addition to your position coordinates (x, y, z). We'll need four equations to solve for the four unknowns. That means we need a minimum of four satellites in view to measure pseudo-ranges from. That's basically the price of using an inexpensive clock. We got off cheap.

Errors in Pseudo-Range Measurements

Pseudo-range measurements taken by a GPS receiver are not perfect. The error is typically less than a half-dozen meters if you are in an open area. Actually, it seems astounding that you can measure the distance to a satellite 20 000 kilometers away with a one-dollar receiver and obtain that kind of accuracy. But accuracy is said to be addictive and let's see how to improve on it.

Multiple sources contribute to errors in pseudo-range measurements. Here are the main culprits.

- Errors in the predicted satellite position and in satellite clock synchronization: You can blame these on the U.S. Government, if you are so inclined, or on the Control Segment, which is charged with operating GPS. The combined error from these two sources has been brought down over the past 20 years from several meters to sub-meter level, an extraordinary achievement.

- Uncertainty in the speed of a radio signal as it travels through the earth's atmosphere: It's not GPS' fault – that's just how the world works. While a radio signal travels in space in a straight line at the speed of light, the last 1000 kilometers of its journey is through variable environments of charged particles (ionosphere) and neutral, gaseous atmosphere (troposphere), where it slows down. The extent of the slow-down depends upon the composition of the atmosphere, which is always changing. This error can amount to several meters.

- Multipath: Closer to the receiver, a signal may reach the antenna through multiple paths – the direct signal from a satellite as well as its reflections from nearby surfaces. These signals interfere with each other, sometimes constructively, sometimes destructively, and the resulting error in pseudo-range can be tens of meters. Fancy receivers use special antennas and processing steps to mitigate this error.

  If you are walking or driving down an urban street with tall glass and steel buildings on both sides, your receiver might not even see a direct signal and only have several versions of reflected signals to work with. The resulting error may be large enough to put you on a different street, and the receiver would typically look for help from an odometer, Wi-Fi data base, inertial instruments, and a street map.

The pseudo-range measurement errors due to the first two sources above are similar for GPS users who are within tens of kilometers of each other. It's a simple consequence of the fact that the satellites are far away and signal paths to two users separated by 10 kilometers are virtually identical.

If you are in a clear space not afflicted with multipath and can somehow determine that your pseudo-range measurement from Satellite #1 is 2.5 meters too short and from Satellite #2 3.5 meters too long, this information would be valuable to another user in the neighborhood. Why don't you be nice and get this information to her quickly, say, on a radio link, or even by posting it on the Internet quickly. She'd be grateful. The errors change slowly with time and you'd have to find a way to transmit the corrections every minute, or so. You can cut down the size of pseudo-range errors for everyone in the neighborhood to a couple of meters, except for the multipath. This approach to correcting GPS measurements has a name: *differential GPS*.

A cell tower can provide assisted-GPS service as well as differential GPS service to cellphones.

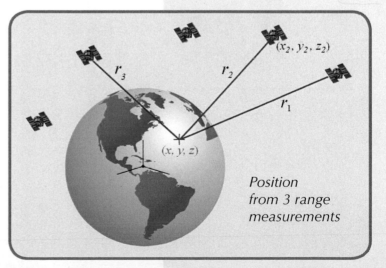

Computing Your Position: Math of GPS

Time for a little math, if you are up to it.

We discuss in this chapter how to compute your position, given (i) the measurements of transit times of GPS signals generated by the receiver, and (ii) the satellite positions at signal transmission time, broadcast by the satellites themselves. We take on some simple, intuitive ideas from geometry, algebra, calculus, statistics, everything. As you'll see, the level of math certainly wasn't the reason we had to wait for GPS until 1995.

The main idea is to solve a set of simultaneous equations in the three coordinates of your position: (x, y, z). We start with the simplest formulation of the problem and add complexity in steps. At the end of each step we'll have a set of equations to solve and you can quit when you think you got what you wanted.

Here are the four steps:

- I. Position from 3 range measurements
- II. Position and time from 4 pseudo-range measurements
- III. Position and time from n pseudo-range measurements, $n > 4$
- IV. Measurement errors and satellite geometry

I. Position from 3 Range Measurements

So, you have 3 satellites in view and your receiver measures ranges to them, which we represent as r1, r_2, and r_3. The corresponding satellite positions are given by their Cartesian coordinates as (x_1, y_1, z_1), (x_2, y_2, z_2), and (x_3, y_3, z_3), respectively. Your position, represented as (x, y, z), is to be determined.

Recalling from Chapter 6 the expression for distance between two points whose Cartesian coordinates are given, we write the measured ranges in terms of the given position coordinates of the satellites and the unknown coordinates of the receiver, giving us 3 equations:

Position from 3 range measurements

$$\sqrt{(x_1 - x)^2 + (y_1 - y)^2 + (z_1 - z)^2} = r_1$$

$$\sqrt{(x_2 - x)^2 + (y_2 - y)^2 + (z_2 - z)^2} = r_2$$

$$\sqrt{(x_3 - x)^2 + (y_3 - y)^2 + (z_3 - z)^2} = r_3$$

(10.1)

We have 3 equations in 3 unknowns: (x, y, z), your coordinates.

We learned to solve *linear* simultaneous equations with paper and pencil in high school algebra, but these equations are nonlinear (actually, quadratic) and it would take a little more work to solve them. If you were feeling ambitious, you could use the Newton-Raphson method to solve these equations iteratively to converge to a solution or you could use a function from MATLAB®. We are all set, in principle—solve the 3 equations in the 3 unknowns, and you have your position coordinates.

That was an idealized case. Recall that the receiver clock will have to be synchronized with satellite clocks if you wanted to measure ranges, but you are too smart to tie up thousands of dollars unnecessarily in a receiver clock. So, let's reformulate the equations for a more realistic case of a $2 crystal oscillator.

II. Position and Time from 4 Pseudo-range Measurements

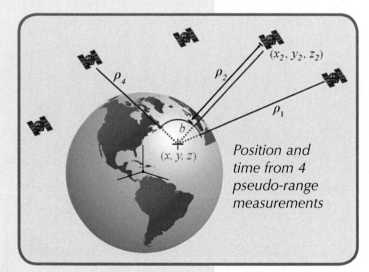

Position and time from 4 pseudo-range measurements

If your receiver clock is off by a certain unknown amount from the satellite clocks, it would affect all your range measurements equally: all would be too short, or too long, by the same amount, which would change as your clock drifts. And you remember from high school algebra: another unknown, another equation.

So, we now have 4 unknowns: the 3 coordinates of your position (x, y, z), and the current receiver clock bias, which we represent as b. To make things easy, we express b in meters. That's just the receiver clock bias in seconds multiplied by the speed of light. We need pseudo-range measurements from 4 satellites in order to set up the 4 equations in 4 unknowns.

The location of the fourth satellite is given as (x_4, y_4, z_4) and we now represent the measured pseudo-ranges using the Greek symbol *rho* as $\rho_1, \rho_2, \rho_3,$ and ρ_4. The equations now are:

$$\sqrt{(x_1-x)^2+(y_1-y)^2+(z_1-z)^2}-b=\rho_1$$
$$\sqrt{(x_2-x)^2+(y_2-y)^2+(z_2-z)^2}-b=\rho_2$$
$$\sqrt{(x_3-x)^2+(y_3-y)^2+(z_3-z)^2}-b=\rho_3$$
$$\sqrt{(x_4-x)^2+(y_4-y)^2+(z_4-z)^2}-b=\rho_4 \qquad (10.2)$$

Solve the 4 equations in 4 unknowns and you'd know your position and time.

We have pretended so far that we knew the satellite positions perfectly and our measurements of pseudo-ranges were error-free. We would have titled the last two sections as 'Positions from 3 *Perfect* Range Measurements' and 'Position and Time from 4 *Perfect* Pseudo-range Measurements' if that didn't sound so pedantic.

In the idealized world of perfect knowledge, measurements from a 5th satellite would bring us no benefit. But in the real world, redundant measurements can improve the accuracy of your position. We take that up next.

III. Position and Time from *n* Pseudo-range Measurements, *n* > 4

If you have only 4 satellites in view, giving 4 pseudo-range measurements, there is no option but to continue to pretend that our measurements are perfect and take a hit in the accuracy of our computed position. But if you have 5 or more satellites in view, giving more measurements than we absolutely need, we can leverage the extra measurements to average out the errors in some sense.

GPS has maintained a constellation of about 30 satellites over the past decade and most users with an unobstructed view of the sky see 8-10 satellites, giving them 8-10 pseudo-range measurements to work with.

Suppose you have *n* pseudo-range measurements, *n* > 4. Continuing with our notation for pseudo-ranges and satellite positions, we rewrite the *n* simultaneous equations like (10.2), but replace the *equals* sign (=) with *approximately equals* sign ($\approx$).

$$\sqrt{(x_1-x)^2+(y_1-y)^2+(z_1-z)^2}-b\approx\rho_1$$
$$\sqrt{(x_2-x)^2+(y_2-y)^2+(z_2-z)^2}-b\approx\rho_2$$
$$\vdots$$
$$\sqrt{(x_n-x)^2+(y_n-y)^2+(z_n-z)^2}-b\approx\rho_n \qquad (10.3)$$

We now have *n* approximate equations in 4 unknowns. That was the easy part. The hard part is to define what we mean by a solution to these approximate equations.

There is an infinite number of values of (x, y, z) and b which may appear to satisfy (10.3). If we take 4 of the approximate equations at a time and solve them as equations, we'll get different answers. We have to identify a solution that fits the complete set of approximate equations 'best' in some sense. Fortunately, we don't have to invent anything. Karl Friedrich Gauss, an iconic figure from the 19th century in the world of mathematics, left us with ideas that we are still plowing. One basic idea is called least-squares.

For a specific set (x, y, z) and b, let's define a *residual* as the disagreement between the values on the two sides of $\approx$ in each of the relations (10.3). We'll define the 'best' solution as the values (x, y, z) and b for which the sum of squared residuals is the smallest. We'll call the solution the *least-squares solution*.

We'll formulate the problem as one of minimization. We'll say the best solution to (10.3) consists of those values of (x, y, z) and b that minimize the value of the following expression:

$$\left(\rho_1 - \sqrt{(x_1 - x)^2 + (y_1 - y)^2 + (z_1 - z)^2} + b\right)^2 +$$
$$\left(\rho_2 - \sqrt{(x_2 - x)^2 + (y_2 - y)^2 + (z_2 - z)^2} + b\right)^2 +$$
$$\dots \qquad\qquad\qquad\qquad\qquad\qquad +$$
$$\left(\rho_n - \sqrt{(x_n - x)^2 + (y_n - y)^2 + (z_n - z)^2} + b\right)^2 \qquad \text{(10.4)}$$

That looks messy, but it's really a simple function of 4 independent variables, (x, y, z) and b.

You may remember from calculus that a smooth function like (10.4) achieves its minimum or maximum value at a point where its partial derivatives with respect to each of the independent variables is zero. So, take the partial derivatives with respect to x, y, z, and b, and set them to zero. We are back to 4 equations in 4 unknowns. Solve the equations, and you have your position and time. Actually, now we should say *estimates* of position and time.

What can we say about the accuracy of these estimates? We can certainly say that the better the agreement between the two sides of the 'approximately equals' sign in (10.3), the better the quality of our estimates, but that's only half of it. We look at the other half below.

IV. Measurement Errors and Satellite Geometry

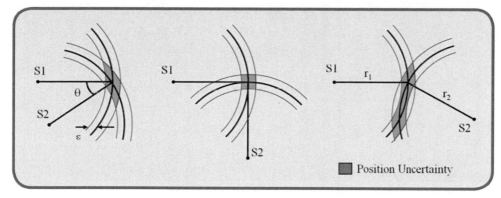

Position Uncertainty

The quality of a position estimate depends upon both the quality of the range measurements (ε) and the geometry of the ranging sources (θ). The role of the geometry is highlighted by keeping the range measurement error the same. The shaded regions represent the resultant positioning uncertainty in three cases.

You still with me? Here's your reward—a subtle idea, explored geometrically.

We still have n pseudo-range measurements, $n > 4$, and our equations are as in (10.3).

If your pseudo-range measurements were perfect, it wouldn't matter where in the sky the 4 satellites were. But the effect of errors in the measurements can get blown up, or can be contained, depending upon where the satellites are located in relation to you.

The spatial distribution (or geometry) of the satellites is often characterized in terms of a parameter called dilution of precision. The more favorable is the geometry, the lower is the value of this parameter. We'll explore the idea in 2-dimensions—positioning on a plane. Measurement of ranges to two radio signal sources would nail down your position.

In the figure above, S1 and S2 serve the role of satellites. We measure ranges, but not without error. Measurement error, shown as ε, creates an area of uncertainty in your position, shown shaded. The role of the geometry is highlighted by keeping the range measurement error (ε) the same in the three cases. We move the 'satellites' around to see the effect on the region of uncertainty. The geometry of the signal sources is optimum for the middle case, as indicated by the smallest shaded area.

For best three-dimensional positioning, a user would be surrounded by satellites, several of which would be low in the sky and some high. If you had only four satellites in view, the ideal geometry would be for one to be directly overhead and three distributed evenly 120 degrees apart just above the horizon. With a 30-satellite GPS constellation, the problem of poor geometry generally arises because the user can't see a significant part of the sky.

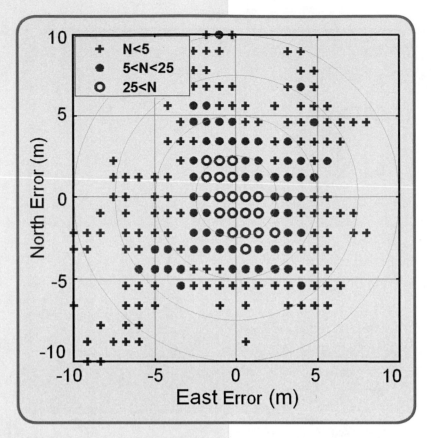

Position estimates obtained 1-minute apart over a day from a stationary, consumer-grade GPS receiver in a clear area (N: number of samples in a cell). The error is no worse than a couple of meters much of the time.

Pseudo-range measurement errors and satellite geometry change over time as satellites move in the sky. That means even if you were to hold still, your estimated position computed by your receiver wouldn't be constant. If you are in an open area, chances are your position estimates would be mostly within a couple of meters of each other. If you are in an urban canyon with a limited view of the sky and multipath galore, there is no telling how bad your position estimate might be. The receiver display would try to keep you on the sidewalk if you are walking, and in the correct lane if driving, up to a point.

GPS and Relativity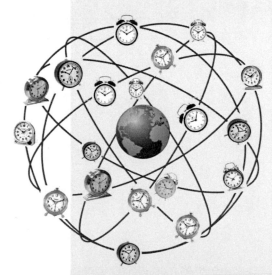

Newton's laws of motion meet the needs of engineers.

So I had concluded as a college student. Einstein's theory of relativity seemed too subtle, too abstract, and too mathematical. Besides, it had no consequences for everyday life.

The familiar concepts of length, mass, and time become nebulous at speeds approaching that of light. But nothing with real mass, like a toaster or a train, is a candidate for the kind of speed that'd require us to take into account time dilation predicted by special relativity. Similarly, studies of black holes and gravity waves may need to account for general relativity, but no real engineering systems need to worry about the unimaginably minuscule differences in time among clocks. Besides, there were no clocks that kept time with accuracy and precision of nanoseconds (i.e., billionths of a second), or better, in those ancient times. So, what if you mistimed an event by a few nanoseconds? In short, engineers didn't have to learn about relativity. That's almost true; GPS is an exception.

It is timely to remind ourselves that advances in space-qualified atomic clocks were a key development that made GPS possible. Turns out GPS must take into account both special relativity and general relativity in order to deliver precise position and time. So, let's talk about the problems that would have arisen if the engineers had ignored relativity in their design of GPS.

GPS is a Bunch of Synchronized, Near-Perfect Clocks in Orbit

It's a mantra worth repeating.

In order to measure ranges to GPS satellites with meter-level accuracy, the clocks aboard the satellites have to keep time with nanosecond-level accuracy.

As we have discussed, the clocks aboard GPS satellites are extraordinarily stable, typically to one part in 10^{13} over a day, which is another way saying that they could gain or lose on average 10^{-8} seconds, or 10 nanoseconds, over 10^5 seconds, which is roughly the length of a day. (A day is actually 86,400-seconds long.)

So, if we can upload clock corrections to each satellite once per day,

GPS is basically a bunch of synchronized clocks in orbit

the error in the transmission time stamped on the radio signal may grow on average to about 10 nanoseconds between the daily updates. That's an error of about 3 meters in range computation and, speaking roughly, an error of about 3 meters in the position computed by the receiver. We could be happy with that.

Gravitational and Motional Effects on GPS Clocks

Our previous calculation of the timekeeping error of a satellite clock would have been fine, but we overlooked an important fact: We pretended as though the clocks were at rest on the earth at mean sea level. So, let's see what relativity has to say about clocks in 20,000-kilometer high circular orbits around the earth. These clocks move at a speed of about 4 kilometers per second and exist in an environment where earth's gravity is only one-fourth that at sea level.

According to the theory of special relativity, a moving clock ticks more slowly when compared with one that's stationary at sea level. A clock aboard a GPS satellite, moving at a rate of about 4 kilometers per second, will <u>lose</u> about 7 microseconds (i.e., seven millionths of a second) per day.

According to the theory of general relativity, a clock in a weaker gravitational field will tick faster than one that's stationary at sea level. Apparently, gravity weighs down time, too. A clock aboard a GPS satellite in a medium earth orbit will <u>gain</u> about 45 microseconds per day over a clock that's at sea level on the earth.

Relativistic Frequency Shifts for GPS Clocks

- **Special relativity**

$$\frac{\Delta f}{f} = -\frac{v^2}{2c^2}$$

$$\simeq -\frac{1}{2}\left(\frac{4\cdot10^3}{3\cdot10^8}\right)^2 \simeq -1\cdot10^{-10}$$

- **General relativity**

$$\frac{\Delta f}{f} = \frac{\Delta\Phi}{c^2}$$

$$\Delta\Phi \simeq \frac{GM}{r_E} - \frac{GM}{4r_E} = \frac{3}{4}\frac{GM}{r_E} \simeq \frac{3}{4}\cdot\frac{4\cdot10^{14}}{6.4\cdot10^6} \simeq 5\cdot10^7$$

$$\frac{\Delta f}{f} = \frac{\Delta\Phi}{c^2} \simeq \frac{5\cdot10^7}{9\cdot10^{16}} \simeq 5\times10^{-10}$$

$\frac{\Delta f}{f}$: fractional frequency offset of an oscillator (s/s)

v : speed of the platform (4,000 m/s)

c : speed of light (3 x 10^8 m/s)

r_E : radius of the earth (6.4 x 10^6 m)

$\Delta\Phi$: change in gravitational potential (m^2/s^2)

Net effect: A GPS satellite clock gains about 38 microseconds per day. This effect is secular, meaning the time offset will accumulate and grow from day to day.

As an interesting aside, note that the effects predicted by special relativity and general relativity cancel each other for clocks located at sea level anywhere on earth. Consider two clocks, one located at the North or South Pole, and the other at the equator. The clock at the equator would tick slower because of its relative speed due to earth's spin, but faster because of its greater distance from the earth's center of mass (about 22 kilometers, as pointed out earlier) due to the flattening of the earth. Because the earth's spin rate determines its shape, the two effects are not independent, and it's no coincidence that they cancel exactly.

What if GPS Forgot about Relativity?

What would have happened if the engineers responsible for designing GPS had disregarded relativity? If the GPS satellites were in fact in identical, circular orbits, their clocks would have shown a puzzling, but identical, behavior of gaining time over clocks of the Control Segment on the earth at a steady rate, about 38 microseconds over a day, the combined effect of special and general relativity.

What would that do to range measurements? A GPS receiver would have measured the ranges to all satellites in view as too short by a common amount (up to about 11 kilometers between daily uploads of clock corrections). What would that do to the position solutions? Turns out, the net effect on position computation would have been exactly zero. As we saw in the last chapter, a common error in the range measurements is harmless in our pseudo-range-based position estimation.

But what if you were looking to GPS to synchronize clocks and keep time on the earth? Clearly, the timing accuracy would have suffered to the extent of 38 microseconds per day between updates of the clock parameters. The systematic errors, once recognized, could have been accounted for by engineers in many ways. But, why not do it right and keep the finicky physicists happy.

The relativistic effects discussed so far can be compensated for easily by setting the frequency of the satellite clocks lower (by 0.00457 hertz) in what's called 'factory offset': The frequency of a satellite clock is set to 10.22999999543 megahertz so that it will tick in orbit at the same rate as a 10.23 megahertz atomic standard at sea level on the earth.

What an ingenious solution!

But What about Eccentric Orbits?

Yes, that's a complication.

The factory offset of the frequency would have taken care of the relativistic effects completely if the GPS satellite orbits were perfectly circular and identical. They are not. You can't control an orbit perfectly.

Each orbit is distinct and slightly elliptical. A consequence of this is that the satellite speed is not constant: The farther away a satellite gets from the earth in its elliptical orbit, the slower it moves; and the farther away the satellite, the lower is the gravity field. That means the clocks in different satellites are speeding up and slowing down at different times and at different rates. The effect for each clock is periodic and quasi-sinusoidal. Averaging the effect over an orbit, we get zero.

For a satellite in an orbit with an eccentricity of 0.02, the net effect is that a clock can be ahead or behind by as much as 45 nanoseconds. The corresponding range error amounts to ± 15 meters. Engineers would have found

> **What's the net relativistic effect on a GPS satellite clock?**
> It can add up to about 38 microseconds over a day.
>
> That may not seem like much if you are running to catch a bus, but a radio signal travels 12 kilometers in that time.

a way to cancel this effect in relative positioning, but there is now no graceful exit. This effect must be accounted for specifically for each orbit, and it is, in the receiver, based on the orbital parameters broadcast by each satellite giving its position and velocity.

There is more to relativity than the special theory and general theory. There is the Sagnac effect associated with our rotating reference frames attached to the earth in which we'd like to determine a position. The principle of constancy of the speed of light cannot be applied in a rotating reference frame, where the paths of the radio rays are not straight lines, but spirals. (Receivers at rest on the earth are actually moving quite rapidly: about 465 meters per second at the equator.) There is also the Shapiro delay associated with the slowing of electromagnetic waves as they near the earth, which amounts to a fraction of a nanosecond. Having mentioned these subtleties, we'll move on.

Does GPS Validate Einstein's Theory?

Yes, if another proof were needed.

What seems more interesting, could Einstein have imagined one hundred years ago that a billion people will unknowingly account for the effects of relativity in their everyday activities?

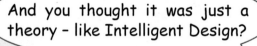

Einstein sketch from ClipArtBest.com

Location Awareness: What's Around Me? Where is He, She, or It?

Location awareness—it's a whole new concept.

It takes a combination of instant location from GPS with access to data bases and wireless communication to take you from *'where am I?'* to *'what's going on around me right now?'* It's about being placed at the center of a map. And the map moves as you do, keeping you at its center. You are important. It's all about you.

There are movie houses, restaurants, and shops around you, all waiting for your business. Click on the movie house icon and you get a list of the movies being shown and their show times. Click on the restaurant and you can see what's on the menu and how long is the wait for a table. Well, you get the idea.

By opening an app like Foursquare or Loopt on your smartphone, you can broadcast your location and see where your friends are right now. Businesses can offer you free products as you walk by. Apps like Yelp, OpenTable, and Urbanspoon offer access to huge data-bases of restaurants, shops, hotels, doctors, art galleries and museums, complete with phone numbers, directions, and customer reviews.

http://www.funmobility.com

With a few strokes on your smartphone you can see where your loved ones are—spouse, children, pets, aging parents. If you are an employer with a fleet of trucks on the road, or you are the director of emergency services for a city with a fleet of ambulances, you'd want a more elaborate display, perhaps covering a wall in your office. If you are a probation officer, you can quickly determine if any of your charges has violated his curfew.

Then there is location-aware advertising. What's the point of a taxi carrying an ad for a restaurant across town, or advertising dinner specials at breakfast time? Why not show today's specials at a restaurant in its neighborhood at the appropriate time? The ads change as the taxi moves around town. That's an example of location-based services, a hot area right now.

And there is location-aware security: A briefcase or laptop that can be opened only in a specified location to minimize the risk of sensitive commercial or military material falling into the wrong hands. Making a big credit card purchase? The merchant can make sure it's you by verifying unobtrusively that the location shown on the cellphone in your pocket matches the location of the shop.

Tracking

Your GPS receiver tells *you* where you are. If others want to know where you are, they will have to sneak a GPS receiver onto your person and have your coordinates transmitted to them somehow, perhaps in real time on a separate radio communication link. That's tracking.

GPS-based tracking devices are now common. If you buy a receiver from an outdoor store or an electronics store, you don't have to worry about Big Brother, but if you rent a car or a boat with GPS, there is a good chance that the rental company can track you or at least record your time-stamped position and speed to determine afterwards where you have been and if you exceeded any speed limits.

Of course, you can choose to be tracked by broadcasting your position to a select group by subscribing to such a service. If you are a policeman or ambulance driver, you may have no choice. Thousands of police cars, ambulances, trucks, and taxis are tracked every day from their headquarters.

Tracking Loved Ones: Where's Jane?

GPS locators are getting small enough to attach to your dog's collar, your child's backpack, or an aging parent's car bumper. You can look up trackers online, either on the Web or through a smartphone app. They can also be configured to send alerts based on various criteria of your choosing. If your teenager's vehicle broke the speed limit, or your dog wandered out of your yard, you get a text alert. You can also look at the bread-crumb trail to see where the tracker has been.

You now have options for secretly tracking the movements of an ex-spouse or employee, though you'll have to be careful when retrieving the unit to re-charge the tracker every few days. You can set up the system to notify you by email or text message when the battery is dying or the unit has been discovered and switched off. You can define zones or addresses that generate automatic alerts: "Jane has arrived at the apartment of a former college boyfriend." (The rules of tracking vehicles vary by state. It's generally considered legal if you're tracking a car that you own, but you don't want to take legal advice from an engineer.)

Cellphone companies offer location-sharing services. It seems strange that anyone would want to broadcast such information all the time, but a cellphone can broadcast such information which can be received by any web-enabled phone or PC. Say you are headed to a meeting and want someone to know where you are until you get there. You open the app and enter the

"Self-Contained, Weatherproof, Motion-Activated, Magnetic-Mount, Battery-Powered GPS Tracker provides up to 80 hours of motion-activated (no battery drain when vehicle stationary) GPS logging on 2 AA alkaline batteries"
http://thespystore.com/ gps-surveillance-equipment/

phone number or email address of the person with whom you'd like to share this information and the length of time the recipient can track you.

Almost every high-value load on a truck or trailer is protected by a GPS-based tracking system. So, too, are luxury vehicles that make attractive targets for thieves.

Jane Under a Privacy Bubble

So, you fear a jealous ex-boyfriend may be tracking your car with a hidden GPS tracker. Or you *know* that your employer is tracking you but you must make a detour from the prescribed route. What are you to do?

Well, technology can help you out. The GPS signals, as we now have said a few times, are extraordinarily weak and can be drowned out by a competing radio signal that *you* can transmit to disable the GPS receiver in the tracker foisted upon you. You can't be 'seen.'

You know that you can't jam radio signals being transmitted under Government authorization, but an underground industry appears to have grown around GPS jammers. Google 'GPS jammer' and you'd get thousands of links to peddlers of cheap ($30) GPS-denying electronics that plug into what used to be called cigarette lighters inside cars and trucks. These jammers are euphemistically known as personal privacy devices.

Unless you are a police officer or an ambulance driver on the job, you do have a right to privacy and might get away with using these devices—if your privacy bubble doesn't stretch much beyond your car. But you can't be sure. These illegal and unregulated devices without any quality control can jam GPS for others around you and even create a public hazard.

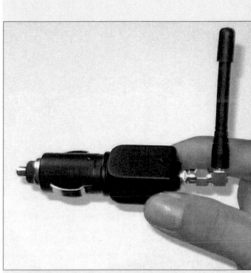

A GPS blocker transmits a low-power signal in the GPS band. It's illegal and can be dangerous to others.

GPS as Big Brother

What's a reasonable expectation of privacy in the age of cellphones and GPS-enabled social networking tools?

The Fourth Amendment to the U.S Constitution:
The right of the people to be secure in their persons, houses, papers, and effects, against unreasonable searches and seizures, shall not be violated, and no Warrants shall issue, but upon probable cause, supported by Oath or affirmation, and particularly describing the place to be searched, and the persons or things to be seized.

Question: Is sticking a GPS tracker to your car a "search"?

The Fourth Amendment to the U.S. Constitution deals with privacy rights of a citizen, guaranteeing the "right of the people to be secure in their persons, houses, papers, and effects, against unreasonable searches and seizures." That means the authorities cannot beat down a citizen's door and rummage around for evidence of a crime. Instead, they are required to present evidence of "probable cause" to a judge who decides whether to allow a search.

But it comes down to what's meant by 'search.'

The challenge lies in applying an 18th century legal concept to 21st century technology. Does attaching a GPS tracker by a magnet to the underside of a car parked on a city street constitute search within the meaning of the Fourth Amendment?

The idea of tracking a suspect's vehicle seems irresistible to law enforcement authorities, especially in cases of narcotics and terrorism. Through a GPS tracker, the police can collect information about who your friends are, where you like to eat, what school your children attend, your hobbies, which doctor you go to, and much more. A case involving long-term GPS vehicle surveillance of a suspected drug dealer was decided by the U.S. Supreme Court in 2012.

In 2007, the police suspected Antoine Jones, an owner of a Washington, DC, nightclub of being a part of a cocaine-selling operation. They placed a GPS tracking device on his Jeep Grand Cherokee without a valid warrant, tracked his movements for a month and used the evidence they gathered to convict him of conspiring to sell cocaine. He was found guilty of drug dealing and was sentenced to life in prison. The prosecutors argued that Jones was driving on public highways and, thus, the police did not need a warrant. But the U.S. Court of Appeals for the District of Columbia Circuit overturned his conviction. The government appealed to the Supreme Court.

In United States v. Jones, the Supreme Court held that placement of a hidden tracking device on a suspect's car constitutes a search within the meaning of the Fourth Amendment. "We need not identify with precision the point at which the tracking of this vehicle became a search, for the line was surely crossed before the four-week mark," Justice Alito wrote. We should note that when the case was argued, a lawyer for the government said the number of times the federal authorities used GPS devices to track suspects was "in the low thousands annually." The case against Jones was thrown out.

"Physical intrusion is now unnecessary to many forms of surveillance," Justice Sotomayor wrote. She added that "it may be necessary to reconsider the premise that an individual has no reasonable expectation of privacy in information voluntarily disclosed to third parties." This so-called third-party doctrine refers to the government getting hold of any information you give to someone else, like a wireless service provider, who knows your location all the time, not with the meter-level accuracy of GPS, but within hundreds of meters in the area covered by a cell tower. And data bases at Amazon,

Google, and Facebook are packed with information on what you like, down to the size and style of your underwear.

There is surely more to come on the privacy issue.

Tracking Unloved Ones

Over two million Americans were incarcerated in 2013, some of them ex-governors of Illinois and Louisiana elected on tough-on-crime platforms. The U.S. has 4% of the world's population, and 25% of the prison population. It's expensive and the governments are broke. It's not good for public safety and it's not good for public spending.

We have to re-think the whole incarceration-rehabilitation thing. I remember the Reverend Jesse Jackson, who in his prime had the gift of rhyme, making up a ditty about how it costs less to send a kid to Harvard for a year than to keep him in jail. He had a point.

When an offender is assigned a GPS monitoring system, a corrections officer locks it into a non-removable ankle bracelet. Offenders also are given a locator box, which tracks the GPS signal and can receive text messages from DOC staff.

Little societal purpose is served by incarcerating nonviolent offenders, like Jesse Jackson, Jr., the former congressman from Chicago, who recently completed a 30-month prison sentence for buying a $43,000 Rolex watch and other luxuries with campaign contributions. We have to find a more constructive way for the likes of Jesse, Jr., to make amends without costing society quite so much.

It's a depressing statistic: According to *The New York Times*, one of 60 people in the United States — almost five million altogether — is either on probation or on parole. The governments are particularly concerned about the 150,000 or so convicted sex offenders and domestic abusers who have been released from prison. About one-quarter of the prisoners serving time for rape or sexual assault were on probation or parole at the time of their latest offense. A victim can get a restraining order but has a hard time proving that the named stalker came on her driveway and slashed her car tires in the middle of the night

Offenders are told to keep the locator box charged. If the battery runs out or the offender strays more than 150 feet from the locator box, the offender's tracking device beeps; community-corrections staff members also receive an e-mail update each day.
from THE SEATTLE TIMES, Jennifer Sullivan, Dec 13, 2007

An idea now catching on is to let GPS watch them—a euphemism for tracking parolees, sex offenders, and stalkers constantly using GPS. It's a simple idea. Put together a box containing a GPS receiver, cell phone connection to law enforcement, and a proximity sensor—a device that sends out a radio pulse every so often to ping an electronic bracelet worn by the person being watched and listens for the response.

The box can sit on a window sill or a car seat with a view of the sky to determine its position. Its position is reported into a central computer every few minutes. If someone tampers with the bracelet or the person being watched moves outside the range of the box, an alert goes out to the authorities immediately. The best thing is that the box can go from home to a workplace, or a store in the neighborhood, as long as it stays in an area designated as 'safe.' If the box moves into an area forbidden for the offender to enter, the authorities are alerted immediately.

It's not prison. It's not freedom. It's a gray area in between, and perhaps the most important innovation in the criminal justice system. It may cost $50 a day to keep someone in the state prison system, and $5 a day to let GPS watch them.

GPS as a Tool of Science

We offer four examples.

The first three, important but straightforward, exploit the basic capabilities of GPS for precise positioning and timing. The fourth is a surprise: using GPS radio signals for remote sensing of temperature and water vapor in the atmosphere for weather prediction and climate change.

How Tall is Mount Everest?

Measuring the height of a Himalayan peak in 1840 couldn't have been easy, but Colonel George Everest, Surveyor General of India, supported by thousands of coolies and hundreds of elephants to haul bulky equipment, managed the task as a part of the Great Trigonometric Survey to measure the prized colony. The peak was tall enough to be recognized as the highest on the earth and was named Mount Everest, in honor of the surveyor—the British are very civilized about that sort of thing.

An aerial view of the southern side of Mount Everest
(Wikimedia photo)

The height of Mount Everest was established in 1852 by the Survey of India (SOI) as 29,002 Indian feet. (Don't ask.) The second determination of the height came a hundred years later, in 1954, from SOI, now under new management. This time the derived height was 29,028 feet (8848 meters) with an uncertainty of ± 10 feet (± 3 meters). The exercise entailed measurements of vertical angles from distant stations two to three hundred miles away and 10 to 12 thousand feet lower than the peak.

Why not just climb up with a GPS receiver and settle the issue?

A survey team sponsored by the U.S. National Geographic Society in 1998 succeeded in collecting GPS data at a number of sites around Mount Everest, but not at the peak itself. The first GPS data collection at the peak occurred for an hour on 5 May 1999. The 1998 and 1999 data sets analyzed together established the height of Mount Everest as 29,035 feet ± 6.5 feet (8850 meters ± 2 meters).

Why so much uncertainty when GPS can nail a position within centimeters? Yes, GPS can indeed nail a position within millimeters if you are happy with (x, y, z) coordinates specified in WGS 84. But height is a different matter. Height relative to what? The traditional reference is the mean sea level, or geoid, an idealized surface covering the globe that has the same gravitational potential as the mean sea level. The uncertainty comes mainly from the definition of the geoid.

There you have it, except for the issue of the changing snow cap and that Mount Everest is growing taller by about 4 millimeters per year due to uplift caused by the Indian tectonic plate pushing northward into Asia.

The Chinese mounted an expedition in 2005 that utilized ice-penetrating radar in conjunction with GPS equipment. The result was what the Chinese called a 'rock height' of 29,017.16 feet (8,844.43 meters), which, though widely reported in the media, was recognized only by China. Nepal, in particular, disputed the Chinese figure, preferring what they termed the 'snow height' of 29,028 feet. In April 2010 China and Nepal agreed to recognize the validity of both figures.

Good move. If you are tiny Nepal, you don't want to mess with China on so sensitive a topic.

From Ocean Currents to Motion of the Earth's Crust

If you are of a certain age, you may remember reading about scientists dropping sealed bottles in the ocean just to see where they end up. One of my earliest scientific memories is about such a bottle with a note inside being found by a fisherman's little girl off the coast of Kerala. She was going to write back in her native Malayalam to the scientists I imagined in white lab coats and thick glasses someplace far away poring over giant maps of the oceans, speculating about currents that would take the bottle from the North Sea to the beach by the little girl's fishing village.

U.S. Department of Commerce
Coast & Geodetic Survey

BREAK THIS BOTTLE

This bottle was released at sea as part of a large-scale study of ocean currents. Information on the date and place of release is on file at the Coast and Geodetic Survey in Washington, D.C.

You can add to the knowledge of ocean currents by returning the addressed card with the requested information on the date and place where you found the bottle. You will receive by return mail information as to where this bottle was released. Your cooperation in giving accurate information will be of great assistance.

Well, take off the lab coats and thick glasses. A GPS receiver inside a clear plastic bubble today can record its entire path—time-stamped position and velocity—and upload data periodically to a satellite overhead, leaving nothing open to speculation and depriving little girls on the Kerala coast of exciting discoveries.

Drop these little plastic bubbles with GPS receivers into a hurricane or typhoon and you'd get a good look at what's going on inside. Drop them into a swollen river and you can monitor the extent of the flooding. Well, you get the idea.

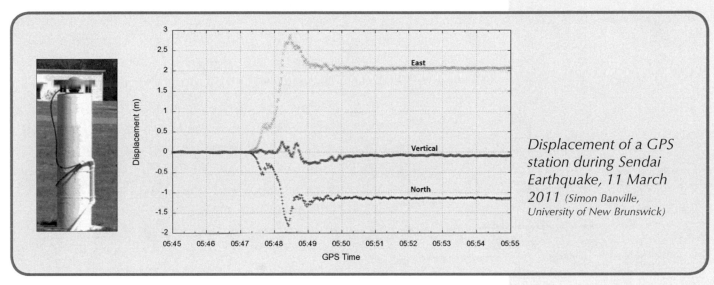

Displacement of a GPS station during Sendai Earthquake, 11 March 2011 (Simon Banville, University of New Brunswick)

Let's switch now from things that move rapidly to things that don't seem to move at all, mostly. Think of the earth's tectonic plates, which move imperceptibly, but relentlessly, up to a point, and then they move cataclysmically. See the plot of the coordinates of the base of a GPS station showing its movement during the Sendai earthquake in Japan in 2011. The ground shook for about 90 seconds and, when it stopped, the base of the GPS antenna had moved by a couple of meters horizontally.

Horizontal velocities due to the motion of the earth's tectonic plates (http://sideshow.jpl.nasa.gov/post/series.html)

Hundreds of GPS stations arranged in a worldwide network now monitor the movements of the earth's crust. Such movement, typically at several centimeters per year, gives geophysicists a powerful tool, though we seem no closer to predicting an earthquake.

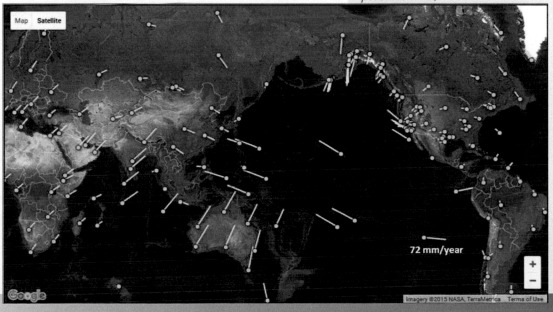

Faster-than-Light Neutrinos?

We briefly recount a recent event (actually, an experiment) that got the normally buttoned-down physicists all excited. The source of their excitement was an announcement from the European Organization for Nuclear Research, a.k.a. CERN, in Switzerland, in the fall of 2011, that subatomic neutrinos were observed traveling a tad faster than the speed of light, in violation of Einstein's special theory of relativity.

The neutrinos are weird subatomic particles. They can go through stone walls, and indeed the earth. They come in three varieties, but can change form as they travel. The experiment was called OPERA, an acronym we'll leave undefined. The idea was to fire a stream of protons from the CERN's accelerator and observe their transformations into neutrinos en route to Laboratori Nazionali del Gran Sasso in Italy.

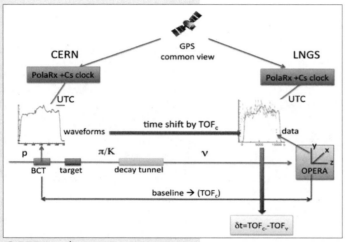

OPERA schematic (CERN graphic)

Dividing the distance between Geneva and Gran Sasso (731.2780 kilometers) by the speed of light (299,792,458 meters per second), the neutrinos shouldn't have reached their destination any faster than 0.002439281 seconds. That's about 2.4 milliseconds. Turns out, the little devils seemed to beat the old record by 60 nanoseconds. Theoretical physicists got very upset. It was like suddenly there was no speed limit.

Maybe there were problems with measurement of distance traveled and/or transit time. For example, a 20-meter error in distance could explain the discrepancy in travel time. We can at least picture an error that size, unlike the 60-nanosecond error (that's 60-billionths of a second) in timing.

So, what's the GPS connection? Well, GPS was the basis for measuring both the distance and travel time. The scientists claimed that they had measured the CERN-Gran Sasso baseline to within 20 centimeters and had measured transit time to within 10 nanoseconds—using GPS.

A basic GPS receiver in your smartphone could measure time to within 100 nanoseconds, if it had to, but we don't need such accuracy in our daily lives. But OPERA needed to synchronize clocks at either end to within 10 nanoseconds. That's a challenge, even with GPS, but doable.

By February 2012 the OPERA physicists had found a couple of flaws with their timing apparatus, one of them a loose wire. In March 2012, the experiment was repeated by a second group. The experimenters clocked the neutrinos racing over the same path from CERN to Gran Sasso going at the speed of light—no faster, no slower. Theoretical and experimental physicists were back on talking terms. We think we know what Einstein would have thought of the fuss they kicked up.

ClipArtBest.com

Monitoring Climate Change

So, we have these two dozen satellites going around the earth twice a day with well-spaced ground tracks and we can measure precisely how long a signal takes to travel from a satellite to a receiver. As we have said, radio signals slow down somewhat as they travel through the earth's atmosphere. The extent of this slow-down is related to the composition of the atmosphere. We could, therefore, monitor the atmosphere globally in real time. But that was too quick; let's back up a bit and slow down.

The travel distance of a GPS signal ranges from about 20,000 kilometers when overhead to about 26,000 kilometers when rising or setting. All but the final 5% of the signal travel can be regarded as in a vacuum or free space, through which the signals travel at the speed of light. Closer to the surface of the earth, at a height of about 1000 kilometers, the signals enter an atmosphere of charged particles, called the ionosphere. Later, at a height of about 50 kilometers, the signals encounter our gaseous atmosphere, made up of dry gases (mostly nitrogen and oxygen), and at about 10 kilometers, gases and water vapor. The atmosphere changes the velocity (i.e., speed and direction) of propagation of radio signals. This phenomenon is referred to as refraction. Change in propagation speed changes the signal transit time, which is the basic measurement of GPS.

The refractive index of a medium is defined as the ratio of the speed of the signal in a vacuum to the speed in the medium. The refractive index of our atmosphere changes along the signal's path, as the density of the air mass changes with pressure and temperature. We are not talking about large differences. The refractive index of the atmospheric gases is about 1.0003 at sea level and even closer to unity as you go higher. The density of the dry gases depends upon latitude, season, and altitude, and is relatively stable. The density of the water vapor can change quickly with local weather.

Water vapor is the engine of the weather and its distribution plays an important role in weather prediction and global climate. GPS gives us a way to estimate water vapor distribution globally from a network of GPS receivers at known locations. Here's how it works. Knowing your location, you know the range to a GPS satellite and can figure out how long the signal would take if it weren't slowed down by the atmosphere. Knowing the signal delay, you isolate the part to be attributed to the water vapor content. Now it's a matter combining these data from the network of GPS receivers for the bigger picture and including these data in an operational way with other meteorological measurements for weather prediction.

The slowing-down of a signal is one aspect of signal refraction; bending of the ray is another. The extent of the bending, measured as refraction angle, like the extent of the slow-down in speed, depends upon the temperature and composition of the atmosphere. And that takes us to GPS potentially serving as a global thermometer.

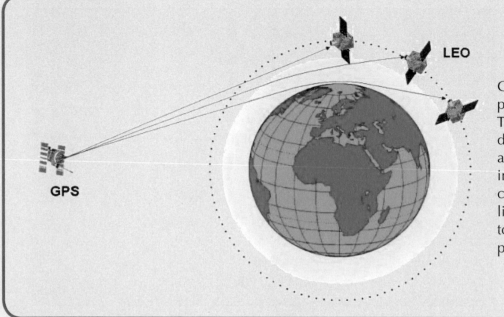

GPS signals bend as they pass through the atmosphere. The amount of bending depends on the temperature and amount of water vapor in the atmosphere. The angle can be measured from satellites in low earth orbit (LEO) to give a global temperature profile. (Image: GeoDAF)

The idea is to exploit a relatively new technique called GPS radio occultation for atmospheric measurements. Occultation means blocking or hiding something from view. If you were riding on a low earth orbit satellite at an altitude of 400 kilometers and looking out the window, you'd see GPS satellites rising and setting over the earth's limb every few minutes. How far below the horizon a GPS satellite is when your receiver loses track of its signal tells you the refraction angle of the atmosphere. (The accompanying figure greatly exaggerates the refraction angle.) You are getting the earth's temperature at macro level.

Scientists have predicted that the refraction angle will increase by 4% in 10 years if the current warming trends continue.

Smart Bombs, Smart Parachutes

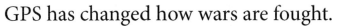

GPS has changed how wars are fought.

A GPS receiver is now found on virtually every U.S. military platform. Every airplane, ship, tank, truck, and soldier receives navigational guidance from GPS with the same assured accuracy as motorists. We take up three examples in this chapter: smart bombs, smart artillery, and smart parachutes. 'Smart' refers mostly to GPS.

Dumb Bombs

For a perspective on delivery of bombs from the air, let's back up a bit and talk about how it used to be done.

Schweinfurt, in the Bavarian region, was a center of German ball bearing production in World War II. Ball bearings don't seem to be a big deal now, but apparently they had such strategic importance that the Allies mounted two large raids on the manufacturing plants—one in August 1943, which proved costly to the U.S. 8th Air Force, and the second in October of the same year, which proved even costlier.

In the second raid, 229 B-17s, known as Flying Fortresses, took off from bases in England, unescorted through much of their mission. Sixty didn't return. Of a crew of 3000, about 600 lost their lives. Millions of pounds of bombs rained down on the town. A thousand civilians were killed and the town was destroyed. But the bombing proved to be only a temporary setback for the German ball bearing production.

Starting with Guernica in 1937, sowing terror from the sky on defenseless populations came to be expected as an unavoidable feature of wars. The U.S. dropped a million pounds of bombs in a single day in 1972 over the port city of Haiphong in North Vietnam. You may remember from TV news how the doors of

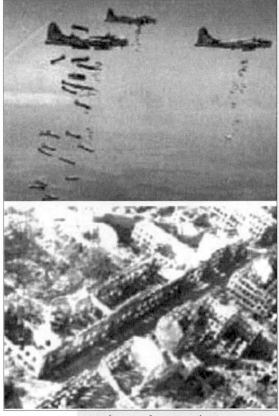

Schwienfurt Raid 1943

the bomb bay would open wide and hundreds of bombs would stream out and fall in a wide swath destroying everything in their path. A pilot would have had a little more control over where the bombs fell if he could fly lower, but that would have exposed him to anti-aircraft fire.

Smart Bombs

The wars would likely go on but the carnage that has come to be known euphemistically as collateral damage would stop. What's changed? Mostly, it's GPS. With technology now available for surgical strikes from the air (and even from cannon-fired artillery) on strategic targets, any indiscriminate terror bombing would open a leader to trial in an international court of law on charges of crimes against humanity.

By the time of the first Gulf War in 1991, U.S. had the first generation of bombs that could be steered to their targets. These were mostly laser-guided bombs, which relied on the target being illuminated by laser. But that meant the aircraft had to stick around to illuminate the target. There was also the problem of bad weather. Sandstorms, dust, and smoke posed a problem in the desert; clouds and rain did the same over the next testing ground in Yugoslavia in 1999. Still, on a clear day, you could destroy a vital bridge without destroying the town.

Then came the GPS-guided, launch-and-leave (or fire-and-forget) bombs. The basic idea is simple: Enter the coordinates of the target and let the control surfaces on the bomb execute a glide to the target, guided by GPS. You could render a military air field unusable by cutting the taxiways and runways precisely at pre-calculated places.

On June 7, 2006, time was up for Abu Musab al-Zarqawi, the leader of Al-Qaida in Mesopotamia and the founding father of the organization that would become the Islamic State (a.k.a. ISIS or ISIL). An F-16C delivered a 500-pound GPS-guided bomb addressed to the house at 33°48′02.83″N, 44°30′48.58″E in the city of Baqubah, where Zarqawi was attending a meeting. The house and its six occupants were killed, but no bystanders were hurt.

This is an ex-airfield!

The strike was carried out by a single B-2 at night after flying from Whiteman AFB nonstop.
(U.S. Air Force photo)

From a dumb bomb to a smart bomb: Joint Direct Attack Munition (JDAM) is a tail kit with adjustable tailfins and a GPS receiver. Strapped on a dumb bomb, JDAM allows it to chart its path to the target.

An F-16 drops a JDAM-equipped GBU-31 2,000-pound bomb (U.S. Air Force photo)

Of course, GPS can't compensate for what's called target location error, or TLE in military parlance. Recall the diplomatic row in 1999 when a GPS-guided bomb landed on a building in Belgrade that would have been a proper target for the U.S. a few months earlier, but was the new home of the Chinese Embassy.

A new example is the Small Diameter Bomb, properly known as GBU-39B SDB I. GBU stands for guided bomb unit. Its small size allows a strike aircraft to carry more of the munitions. The glide wings, which deploy after release, give it a range of about 100 kilometers. With only about 50 lbs of explosives, an SDB makes up for its lack of explosive punch with the precision of the strike: Its circular error probable (CEP) is 5 meters, meaning half the time an SDB would hit within that distance of the intended target. (We'll use the term CEP a couple of more times in this chapter.)

GPS-guided Small-Diameter Bomb (U.S. Air Force graphic)

We should note that the five ball bearing factories of Schweinfurt would be taken out today with five GPS-guided 2000-pound bombs delivered from a single B-52. GPS is called a force enhancer, and now you know why.

Unmanned Aerial Vehicles

UAVs or drones are a new thing.

They wouldn't exist without GPS. GPS receivers on the drones tell their location, speed and attitude to the pilots, who may be sitting in front of computer monitors in a building thousands of kilometers away.

Military drones have been used for a few years for surveillance and targeted killings in faraway places of occupants of a house or passengers in a car. It's particularly suited for war against ragtag cells of terrorist groups. To see how much has changed, consider the U.S. intervention in Libya in 2011. Technically, it was not even a war because no soldiers were involved on the ground. It was all conducted by drones.

Predator UAV carrying a Hellfire-C laser-guided missile (U.S. Air Force photo)

You'll soon see more and more drones in the sky above you, operating in the National Air Space. Some would be searching a wilderness area for lost hikers, or mapping the spread of forest fires. Some would be flying for instant delivery of goods bought on the Internet, or for long-distance selfies.

Smart Artillery

Precision artillery would seem to be a misnomer to anyone who has seen any movie footage from past wars. The recoil of the gun seems to give little indication that the round is actually fired at a specific target. And it gets worse when you learn that the explosive charge in the gun imparts to the round 10,000 g's of acceleration leaving the gun. (Recall that 'g' denotes acceleration due to gravity and equals 9.8 meters per second per second.) There would seem little hope that any delicate electronics, like a GPS receiver, installed on an artillery round for guidance and control, would survive the brutal forces of firing. Well, they can.

Meet Excalibur. It's fired out of a cannon like an artillery shell, but beyond a certain point flies like an airplane, hitting its intended target at 50 kilometers with a CEP of 10 meters. The GPS receiver and guidance computer survive the 10,000 g's just fine. At $30,000 per round, it's a bit pricey right now, considering that a basic artillery shell filled with explosives costs about $30.

But let's get away from killing and talk about saving lives.

Excalibur is a 155 mm extended-range, GPS-guided artillery shell. It can hit a target 50 kilometers away with a CEP of 10 meters. (U.S. Army photo).

Smart Parachutes

Parachute resupply used to be chancy business—you just can't get the winds right. Dropped too high, winds might cause resupplies to miss the drop zone; dropped too low, enemy ground fire might hit the aircraft. We now know how to deal with it.

The U.S. military has developed GPS-guided, steerable parachutes which can hit pre-planned targets within 50 meters from 25,000 feet, greatly reducing the size of the drop zones and allowing airdropping of multiple loads from an aircraft in a single pass.

Here's how it works. Steering lines give the parachute directional control by creating drag on one side, or the other. How and when to tug on the steering lines is determined in real time by an airborne guidance unit, which contains: a GPS receiver to obtain continuously the position and velocity of the load; a battery pack; control software and the necessary hardware to operate the controls.

Thousands of airdrops in Afghanistan have been conducted since 2006, many to bases that cannot be resupplied by other means.

JPADS uses GPS-guided steerable parachutes to deliver air drop bundles. (U.S. Air Force photo)

Who Invented GPS?

That's not an easy question to answer.

It's not like asking who invented the steam engine or light bulb. These early inventions were based on discrete ideas to solve narrow, specific problems tackled mostly by individuals. GPS is based on a collection of ideas developed over a long period by many and brought together in a *system* by a large cast. Besides, as the old saw has it, a success has many fathers, and GPS is a success far beyond the dreams of all who were fortunate enough to be at the right place at the right time and, in some cases, with a bright idea.

We'll address the question, but come at it obliquely as we tell the story of how GPS came to be developed. In the account below, we acknowledge some of those who contributed vital technical ideas for system design or provided the essential organizational savvy to implement the design and see it through the fraught acquisition process of the U.S. Government.

What follows is not to be mistaken for an official history or an impartial account.

It's Bureaucracy Out There

The development of a government-sponsored system like GPS is a lengthy, bureaucratic process. There are real inter-service rivalries at the Department of Defense to deal with. There is also a delicate dance among government agencies that want to see a system realized, but each offers its support with a measure of restraint. You don't want to be seen as too enthusiastic for fear that you'd get stuck with the bill.

Once the program is approved and you are named program manager, you look around for godfathers who'd offer protection to the fledgling program. You also have to go before the Congress to make a case for funds, painting a rosy picture of the potential benefits. You then return year after year to justify why the development must go on despite rising costs, delays, and other difficulties that you claim are only temporary setbacks.

It's no easy task to keep a system which would take 20 years from inception to operational status from suffering irreversible loss of confidence and funding along the way. Add to that a complication of the program manager, an Air Force colonel, 'rotating out' of the job every couple of years.

Delta II carrying GPS IIR-11 mission dedicated to Ivan A. Getting (U.S. Air Force photo)

The necessary continuity and cohesion were provided by a sharp cadre of engineers of The Aerospace Corporation, a so-called Federally Funded R&D Center, founded in Los Angeles in 1960 to support the Air Force technology initiatives. Aerospace engineers sat at the elbow of the colonels all through the design, development, and deployment of the system.

After Transit, What?

Transit, the first satellite-based navigation system, was commissioned in 1964. Its success begat programs in the mid-1960s to develop new, improved navigation systems. Each service had its own program, even programs. The Navy and Air Force needed accurate global navigation systems for their aircraft, but even the Army got in the game with a satellite-based system called SECOR (for sequential collation of range).

The Navy already owned Transit and wanted to build a second-generation of the same. An enterprising engineer at Naval Research Laboratory in DC by the name of Roger L. Easton hit upon an idea of a time-based navigational system that would use range measurements from satellites instead of Doppler shifts. The idea required much more precise and stable clocks in the satellite than what Transit needed and Easton's main contribution was to instigate the development and test of such clocks.

Easton proposed placing the satellites in medium earth orbits and suggested a signal design, something called side-tone ranging. Not wanting to draw negative attention of Transit folks in the office next door, Easton quietly obtained a grant of all of $35,000 form a sympathetic boss and started what came to be called the Timation Program to study the behavior in space of quartz crystal oscillators, the best available at the time. He later switched to rubidium atomic clocks when they became available. Even though his proposal on side-tone ranging signals turned out to be a misfire and the Navy ended up with only a minor role in the development of GPS, we'll call Easton a *Co-designer of GPS* for exploring a novel idea.

The Air Force saw space as an extension of its natural domain and didn't want any interlopers. That's how they perceived the Navy efforts. In the mid-1960s, the Air Force's Space and Missile Organization had a program called 621B, an outgrowth of earlier exploratory programs at The Aerospace Corporation. Project 621B was led by Phillip Diamond. Participants included Alfred Bogen, Lawrence Hagerman, Robert Levinson, Howard Marx, Hideyoshi Nakamura, Arthur Shapiro, Richard Slocum, Peter Soule, and James Woodford.

Project 621B provided some key innovative ideas for GPS. The idea of placing the satellites in geostationary orbits was a clunker, but spread spectrum signals for ranging and stable atomic standards in orbit were winners. Both were proposed in the 1966 report of 621B, written by Woodford and Nakamura. We'll call them and their 621B teammates *Co-designers of GPS*.

Co-designers of GPS (from the top) Roger L. Easton of Naval Research Lab, James B. Woodford and Hideyoshi Nakamura of The Aerospace Corporation. Woodford and Nakamura are stand-ins for the Project 621B team at Aerospace.

Architect of GPS

In 1968, the DoD established the Navigation Satellite Executive Steering Group to examine the best way to address its navigation requirements with a DoD-wide capability.

The Architect and Prime-mover of GPS: Col. Bradford W. Parkinson, Program Manager from 1973 to 1978

"The mission of this Program is to: drop five bombs in the same hole and build a cheap set that navigates (< $10,000), and don't you forget it!"

A factor which had a significant effect on the development of GPS was the streamlining of the acquisition process introduced by David Packard, he of the Hewlett-Packard team of entrepreneurial electrical engineers from Stanford University and the model for garage start-ups in Palo Alto. Packard, who was appointed to the number two position at the Pentagon by President Nixon in 1969, favored joint programs among services to foster greater cooperation, or at least reduce the inter-service bickering. GPS became one of the first examples of a joint program.

The GPS Joint Program Office (JPO, pronounced *jay-poe*) was established in 1973 at the Los Angeles Air Force Base. The Air Force appointed Col. Bradford W. Parkinson as its first Director. Parkinson was a lucky choice. He was a navigator, a graduate of the Naval Academy with a master's degree in engineering from MIT and a Ph.D. from Stanford. He was young, brilliant, cocky, and tough. (Though no longer young, I am happy to report that Parkinson, now 80, looks much younger and has mellowed, but remains as brilliant and tough as ever.)

A professional navigator takes charge of a program to develop a new kind of navigation system—how often does that happen in government? Parkinson assumed the responsibility for synthesizing a design of GPS and realizing the system.

Three key elements of GPS design were to be settled:

- Ranging: One-way or round-trip ranges to satellites?
- Signal design: What kind of signals to transmit for precise ranging and economic use of the radio spectrum?
- Satellite orbits: How many satellites and where to place them in space?

In his term as program director from 1973 to 1978, Parkinson recruited top-flight staff and moved with great skill to build coalitions to put the program on a solid footing.

Co-designers of GPS: (from the top) Maj. Gaylord B. Green (USAF), James J. Spilker of Stanford Telecom, and Charles R. Cahn of Magnavox

Parkinson's first hire was Major Gaylord B. Green, USAF, a Stanford astronautics alumnus, whom he enlisted on the way back to his cubicle after receiving the GPS assignment. Major Green was instrumental in the design of the inclined, 11h 58m-orbits of GPS satellites and went on to lead the JPO in the 1980s as a colonel.

Parkinson contracted the signal design task to two highly regarded signal theory gurus: James J. Spilker, founder of Stanford Telecom, and Charles R. Cahn of Magnavox. Spilker and Cahn thrashed out the design of what we have called the Open Signal. They also designed the military signal, which we have stayed clear of.

We'll call Green, Spilker and Cahn *Co-designers of GPS*.

The GPS team soon grew to dozens and then hundreds as the focus shifted to implementation of the design.

Parkinson adopted as the GPS JPO motto: (i) drop 5 bombs in the same hole and (ii) build a cheap set that navigates (< $10,000), and don't you forget it! It's something the generals could relate to. Fancy talk about '10-meter three-dimensional positioning error (95th percentile)' and '100-nanosecond time transfer capability' wouldn't have meant the same thing to the brass.

The part about five bombs in the same hole had seemed realistic in 1975, but the part about getting the price down to $10,000 was considered too ambitious.

We'll call Parkinson the *Architect and Prime-Mover of GPS*.

Godfathers of GPS

In order for the program to proceed, it had to win approval of the Defense System Acquisition and Review Council. That was the first milestone for Parkinson to achieve. Unfortunately, Parkinson forgot that it was meant to be a joint program and proceeded instead to present the Air Force's proposal 621B for approval. His plan was rejected.

That was to be Parkinson's rare misstep. Chastened, he proceeded to synthesize a system that combined the best aspects of the Navy's Timation and the Air Force's 621B. GPS will employ passive, one-way ranging with atomic clocks aboard satellites. It will use spread spectrum signals. It will deploy a constellation of two dozen satellites in medium earth orbits. That's GPS as we know it today.

The Navy continued its Timation effort beyond 1973, but now under the direction of the GPS JPO. Under this new arrangement, NTS-II, containing two cesium-beam frequency standards and built to the GPS concept, was launched in 1977. It demonstrated frequency stability of 2 parts in 10^{13}, giving a timing error of 20 nanoseconds per day. JPO subsequently subsumed Transit into its fold and appropriated its funding to buy two more satellites. The Navy's development programs associated with Transit and Timation disappeared. There was only GPS.

We'll recount a couple of anecdotes we have heard Parkinson tell.

In a bureaucracy, many can say 'no' to a proposal but few can say 'yes.' An example of Parkinson's savvy is reflected in the name he chose for what had

Godfathers of GPS: (from the top) Lt. Gen. Kenneth W. Schultz (USAF), Commander, AF Space and Missile Systems Organization in the early 1970s; Malcolm Currie, Undersecretary of Research and Engineering for the DoD, 1973-1975; and Ivan A. Getting, Director of The Aerospace Corporation in the 1960s and 70s

started out as the Defense Navigation Satellite System. The names Global Positioning System and Navstar were suggested at different times by higher-ups in the DoD, and Parkinson was happy to oblige in order to build support for the program. The system came be known as the Navstar Global Positioning System. Parkinson was also fortunate to have the full backing of his immediate boss, Lt. Gen. Ken Schultz, an aeronautical engineer and bomber pilot.

The support of Malcolm Currie, a physicist and, as Director, Defense Research & Engineering (DDR&E, in DoD-speak), the number three in the Pentagon hierarchy, also proved vital to development of GPS. As Parkinson tells it, Currie, who was appointed to his position by President Nixon, had to make frequent trips to Los Angeles in 1973-74 to settle his affairs and move the family to DC. He looked for things to discuss at the Los Angeles Air Force Base over these business trips. Eventually, they ran out of things and somebody suggested he talk with the young Col. Parkinson about his ambitious project.

Currie was impressed. GPS, which later came to be called Force Enhancer, was a support program for all services, a fact that didn't endear it to the Air Force brass. It wasn't a weapon system and it certainly wasn't an aircraft. If it weren't for Currie's backing the Air Force generals would have cancelled it in favor of buying a few more fighter jets.

The enthusiastic advocacy of GPS by Ivan A. Getting, an alumnus of the MIT Radiation Lab (the legendary Rad Lab from WWII) and Director of The Aerospace Corporation in the 1960s and 70s, was also seen as crucial to the launching and sustaining of the program.

We'll call Schultz, Currie and Getting *Godfathers of GPS*.

The Start: Catch-22

The GPS program in the 1970s faced a paradoxical situation: satellites couldn't be acquired until the idea had been demonstrated as feasible, and the feasibility couldn't be demonstrated without the satellites.

The JPO came up with an ingenious solution: Test the idea in an inverted test range at Yuma Proving Grounds with signal transmitters on the ground and receivers in the air aboard aircraft. The demonstration was a success and prototype satellites could be specified and ordered.

The first satellite of the first batch, called Block I, built by Rockwell International, was launched in 1978. Ten more followed between 1978 and 1985. This partial constellation not only proved the basic GPS concept but demonstrated the system to be capable of delivering a lot more than was promised.

Col. Bradford W. Parkinson (center) and colleagues in GPS Program Manager's office circa 1975. Recognize the sketch on the chalkboard?
(US Air Force photo)

Surveyors and geodesists, the first serious users, showed the system to be capable of providing an unprecedented centimeter- and millimeter-level accuracy for relative positioning of two points. Specialized surveying receivers would soon come to market, revolutionizing the field.

Rocky Road

The GPS program had its ups and downs in the 1980s. The budget was cut along the way, even 'zeroed out' a couple of times, but the program managed to survive. The Space Shuttle had been planned as the launch vehicle for GPS satellites and the program suffered a serious setback when *Challenger* blew up in 1986 shortly after launch. The program was delayed by a couple of years as it regrouped and settled on a Delta rocket as the launch vehicle.

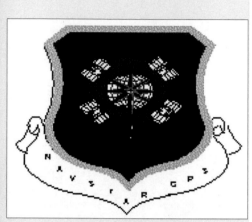

The GPS program had no money for luxuries like a professionally designed logo.

The first operational satellite from the batch of 28 satellites called Block II and Block IIA was launched in 1989. Only a dozen such satellites had been deployed when a big test came with the first Gulf War, called Operation Desert Storm, to expel Saddam Husain from Kuwait in 1991. GPS was a big hit with the troops for simple logistics. A jeep carrying mail or meals could drive straight to the right tent among hundreds pitched in the featureless desert. GPS-guided bombs and artillery rounds were to come a few years later.

Twenty-four satellites were launched between 1989 and 1993 to populate the constellation. The program milestone of Initial Operational Capability was declared in 1993; Final Operational Capability came in 1995.

Modest Expectations and Mega Success

The expectations of the developers of GPS were modest. They couldn't have imagined that GPS would be used by millions every day and be regarded as a utility.

An early Air Force program manager, when asked by the Congress in the early 1980s to estimate the number of potential users of GPS came up with the number 27,000. He must have high-balled the number to keep the program from getting cancelled.

One is reminded of similar prognostications by earlier technology pioneers. Thomas J. Watson, the founder of IBM, is said to have estimated around 1950 the number of computers needed in the world as 12. Ken Olson, the mini-computer pioneer and founder of Digital Equipment Corporation is said to have remarked around 1980 that he couldn't see any value of a computer in a home. Of course, the only application one could think of at the time for a home computer was to keep recipes organized.

For the record, the number of GPS receivers in service in 2015 exceeds 1 billion, most of them built into smartphones.

When pressed on the price of a receiver, the aforementioned GPS program manager expressed optimism that the price would come down in time to $10,000. That, too, had to be a leap of faith because GPS receivers in the market in the early 1980s cost over $150,000. The GPS receivers built into smartphones in 2015 add about $1 to the cost of the phone.

Prizes for GPS

We have named in this chapter a dozen co-designers of GPS, an architect and prime mover, and three god-fathers. As a rule, engineers tend to have little talent for self-promotion, and we have no doubt overlooked some worthy contributions.

Big-name awards for the development of GPS have gone to Easton, Getting, and Parkinson.

Getting and Parkinson shared the Charles Stark Draper Prize of the National Academy of Engineering in 2003 and split $500,000 between them "for their technological achievements in the development of the Global Positioning System (GPS)."

Bradford W. Parkinson (second from the left) and Ivan A. Getting (second from the right) receiving the 2003 Draper Prize for Engineering

The National Academy of Engineering operates under a congressional act of incorporation, signed in 1863 by President Lincoln. Under this charter, the NAE is directed "whenever called upon by any department or agency of the government, to investigate, examine, experiment, and report upon any subject of science or art." Membership of the NAE is about the highest honor an engineer can aspire to. The Draper Prize is often referred to as the Nobel Prize for engineering. It's a big deal. And the NAE credited Getting and Parkinson with the development of GPS.

But a year later, the Bush II White House singled out Easton for the 2004 National Medal of Technology "for his extensive pioneering achievements in spacecraft tracking, navigation and timing technology that led to the development of the NAVSTAR-Global Positioning System (GPS)."

It's interesting that the National Academy of Engineering and the White House saw things so differently, but note that neither used the word 'invention'.

Roger L. Easton receiving the 2004 National Medal of Technology at the White House

So, Whom Do We Thank for GPS?

Yes, perhaps that's the right way to frame the question.

As my friend Per Enge says, start with Euclid, Kepler, Newton, Maxwell, Galois, Abel, Einstein, Rabi, and Clarke. Then there are dozens of hands-on engineers who contributed important ideas. And finally, according to Per, there is Parkinson, now his colleague at Stanford, who was at the right place, at the right time, with a right mix of awesome talents to pull the ideas together and launch the system. Professor Enge's obvious bias aside, he makes an important point.

We'll leave it at that.

Who Controls GPS? 16

That's easy to answer: It's the U.S. Department of Defense.

The DoD will tell you differently. They'll say GPS is controlled by an Inter-Agency board called the National Executive Committee for Space-Based Positioning, Navigation, and Time, which includes not just the DoD, but also the Departments of Agriculture, Commerce, Interior, Homeland Security, State, and Transportation, and NASA.

Yes, that's what it says on paper, but the DoD controls GPS, and it's not a bad thing. The DoD has been an excellent steward of GPS. And, happily, it is now solicitous of civil concerns and requirements. That wasn't always the case.

Who Pays the Piper?

Let's face it: The only reason we have GPS is that the DoD required it.

The DoD planned, designed, developed, and deployed the system and operates it from day to day. The cost of system development was borne by the DoD. The cost of its maintenance and upgrade mostly comes out of the DoD budget. Isn't there a saying—something about paying the piper and calling the tune?

But military ownership of GPS is thought to be bad for public relations and there is a conscious effort to give the story a spin. For FY 2015, the Congress appropriated $1.044 billion for GPS: $1.034 billion went to the DoD and $10 million to the Department of Transportation for the civil-unique capabilities. So, who is paying the piper?

But we don't need to get caught up in PR. It's all paid out of general tax revenues. We don't care about nitpicking as long as GPS is managed competently, and so far it has been—by the DoD.

But let's backup to 1983 and talk about a pivotal event that required the U.S. to think seriously about a policy on who could use GPS.

KAL 007: The Target is Destroyed

On 1 September 1983, Korean Airline flight 007 from New York to Seoul via Anchorage went off-course into Soviet airspace, apparently due to navigation problems. It flew for two hours above super-secret military installations on Kamchatka Peninsula. The Soviet military, preparing for a missile test, appeared to be on edge. It was nighttime and they saw the Boeing 747 as a U.S. spy plane. Two heat-seeking missiles fired from a Sukhoi-15 interceptor shot it down in the Sea of Japan, killing 269 passengers and crew onboard.

The sub-title of this section cites a snatch of the intercepted radio communication between the Sukhoi pilot and his controllers.

The Soviets handled the aftermath in their usual heavy-handed way, first denying any knowledge, and then leveling counter-charges of spying, refusing to release the flight data recorders.

To his great credit, President Reagan stepped up within two weeks of this tragedy to offer GPS, "when it becomes operational in 1988," for aviation worldwide to avert such disasters. GPS took a bit longer, but the U.S. public commitment has stood.

It's My Toy, You Can't Play With It

The DoD at first didn't appear too keen on sharing their shiny new toy with civil users. While GPS was being developed in the 1970s and 1980s, it was understood to be a dual-use technology even though no civil agencies spoke up as enthusiastic backers, leaving the DoD to carry the full weight. As the system approached completion the civil agencies began to see what it could do for them.

The FAA saw the potential to dismantle their billion-dollar infrastructure of VORs, DMEs, and ILSs to guide the airplanes from takeoff to landing. NASA could land the Space Shuttle without investing millions of dollars in a specially instrumented runway. Did they say to DoD "thank you for this wonderful gift"? They didn't. Instead, they sent over a list of what they saw as shortcomings of GPS, to be fixed on DoD's nickel. It wasn't a good start of a relationship.

Sharing GPS with civil users meant sharing it with the Soviet military. Besides, Muammar el-Qaddafi was as crazy in 1990 as he was at his overdue demise in 2011. Why give him this Rolls Royce of a navigation system for aiming missiles at New York?

U.S. Pays to Garbage Up GPS Signals and Then Pays to Clean Them Up?

The DoD adopted a policy called *Selective Availability*. It meant that the signals from each satellite were to be degraded purposefully in a way that the military receivers could undo, but positioning accuracy for the civil users was reduced from several meters to several tens of meters. The resulting positioning uncertainty went from the size of ping-pong tabletop to a football field.

As you would have guessed, Selective Availability was implemented by jiggling (dithering, in technical language) the satellite clocks in a pseudo-random way. Unfortunately, accuracy of 100 meters would have been good enough for Qaddafi's purposes, but drove the civil users in the U.S. nuts.

Within a few years the civil world devised and implemented ways to get around Selective Availability, but it took additional expense. The Coast Guard and FAA built systems that determined the amount of error being introduced in GPS signals and broadcast this information continuously by radio signals for GPS receivers to use.

I vaguely remember a Congressional hearing where a Southern congressman with an excessive drawl and a well cultivated aura of folksiness asked: You are telling me that one part of the U.S. Government pays to garbage up GPS signals and another part pays to clean them up. Have I got that right?

Of course, he had it right and, put that way, hard to defend.

In the mid-1990s, the Cold War was receding in memory and commerce was the new priority. The Clinton-Gore White House reaffirmed the policy first enunciated by Reagan and set in motion a change that led to eventual discontinuance of Selective Availability in 2000. And it's not coming back: The newer satellites have no provision for it.

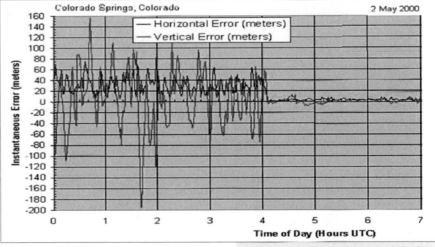

The error in GPS-provided position went down from several tens of meters to several meters as the U.S. policy of selective degradation of the civil signal was discontinued.

(Based on a USAF graphic)

Inter-Agency Governing Board

As we have said, military ownership of GPS was bad public relations.

While civil applications of GPS were growing explosively, there was unease about the military control, even among our allies. You don't want to become too dependent on a system that might be switched off by some military hotheads far away. In order to reassure the worldwide user community, President Clinton issued a Presidential Directive in 1996. Reaffirming the pledge from the Reagan era to maintain GPS signals "for the foreseeable future on a continuous, worldwide basis, and free of direct user fees," it went a step further: It is U.S. policy "to promote integration of GPS into peaceful, civil, commercial and scientific applications worldwide, and advocate acceptance of GPS as an international standard." Clinton assigned the management responsibility to an interagency GPS executive board, co-chaired by representatives of the Departments of Defense and Transportation.

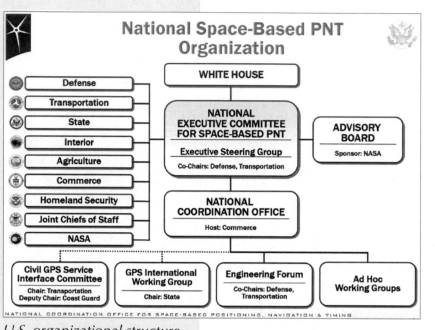

U.S. organizational structure for GPS governance

By 2004, the GPS scene had changed sufficiently for the Bush II White House to issue a new policy directive. While Clinton's directive barely filled three pages, Bush's runs into 11 pages. There is no sharp departure in policy, just more things to address, especially on security. GPS is identified as an element of "U.S. Critical Infrastructure," giving the system special status for funding under U.S. law. The Bush Directive expanded the Interagency GPS Executive Board to include additional agencies of the U.S. Government. The new governing body is called the National Space-Based Positioning, Navigation, and Timing Executive Committee. We referred to it at the beginning of this chapter.

Will GPS be available everywhere at all times?

The short answer is 'no.'

The U.S. Government has laid out its policy in public documents. The USG intends to make the GPS open service at specified performance levels available worldwide, but reserves the right to deny service by jamming signals locally in areas of conflict for reasons of security. It would be prudent for the U.S. to jam GPS civil signals within 100 kilometers of where a Navy ship is discharging troops in an area of conflict.

Beyond the matter of U.S. policy, we also have to be prepared for the inevitable temporary, local breaks in service caused by hackers. Such disruptions of lives and activities, when they come, are likely to be comparable to those caused by natural events like snow storms and floods, which seem to be getting more frequent and severe every year.

While the U.S. would have preferred to have GPS as the sole global navigation satellite system, its unprecedented success made the GPS Wannabes inevitable. More in a bit.

A Day without GPS?

Yes, it *can* happen.

San Diego got a preview in 2007. The disruption started one day just around noon in the harbor area. At the air traffic control tower, the controllers noticed that their system for tracking incoming planes was out of whack. At the Naval Medical Center, emergency pagers used to call doctors stopped working. The harbor's traffic management system had problems. People on the street found they had no cellphone signal. Customers at local ATMs trying to withdraw their money couldn't. It went on for two hours and the mystery wasn't solved for three days.

Turned out, two Navy ships in the harbor were conducting a training exercise. Technicians jammed radio signals to test communication procedures to be used in their absence. Unwittingly, they also jammed GPS signals locally.

GPS as a Utility

In a few short years, GPS has become an essential utility, on a par with water, electric power, telephone service, and the Internet. The GPS-based positioning and navigation service is at the heart of our air, marine and surface transportation capabilities, distribution industries, emergency services, road building, mining, and farming.

In a less visible but even more important role, GPS-provided precise time is the key to digital communication networks and the electricity grid. Cellphone companies use time from GPS to synchronize communications between your phone and the tower, and handoffs between towers. Electricity grids use time from GPS to

JAMMING CELL PHONES AND GPS EQUIPMENT IS AGAINST THE LAW!

Issued by the Enforcement Bureau of the Federal Communications Commission

connect power from energy suppliers so the alternating current connects in phase. Banks and stock exchanges use time from GPS to time-stamp their transactions. ATMs may be doing something similar, given the San Diego experience.

There is concern that our reliance now on GPS may be too great. Let's talk, then, about what it'd mean to us to be deprived of this utility over a wide area for a day. But first a few words on the fragility of this utility and the potential for its widespread loss.

How might we lose GPS? The answer is: through jamming, spoofing, ionospheric storms, and system screw-ups.

GPS signals, as we have now said a few times, are extremely weak—you know, 30-watt transmitters 20,000 kilometers away. The signals are easily drowned out by radio noise, intentional or not. Vulnerability to jamming is not the only problem. GPS radio signals can also be faked. Solar flares can interfere with all radio signals, including those from GPS, potentially denying service over a large area for several days at a time. And, the system, though superbly managed, isn't above screwing up.

Jamming GPS

If you hijack a truck carrying valuable or hazardous cargo, the first thing you'd want to do is to neutralize its GPS-based tracker. The same if you want to take a detour from the prescribed path that your employer might disapprove of.

We have talked previously about $30 jammers, known euphemistically as personal privacy devices that radiate about 10 milliwatts of power centered at the GPS frequency and have a range of tens of meters. These are simple devices, yet highly effective in rendering a GPS receiver within range useless. They are illegal, but readily available. Like most other things, they are manufactured in China. You can also get from the Internet jammer designs for a DIY project.

GPS jammer, blocker, or personal privacy device -- call it what you like -- transmits a low-power signal in the GPS band. It's illegal and can be dangerous to others.

Catching someone in a moving automobile with a 10-milliwatt GPS jammer is not easy, but if you get caught, be prepared to pay a hefty fine. The FCC recently imposed a fine of $32,000 on Gary P. Bojczak, a trucker, who was caught with a GPS jammer on the New Jersey Turnpike that runs by the Newark Liberty International Airport (EWR).

Turns out the FAA was testing equipment for GPS-guided precision approaches at EWR and observing episodes of intermittent jamming of GPS signals. You don't want to be on an airplane executing an approach in poor visibility when the GPS signals are jammed, albeit inadvertently, by some clown trying to hide from his boss.

The authorities had been monitoring that stretch of the Turnpike and nabbed Bojczak as he sped by in his Ford F-150 pickup truck. With dozens of short jamming episodes observed at EWR nearly every day over three years, Bojczak was the sole perpetrator caught by law enforcement. The fine wasn't Bojczak's only problem. His employer, a trucking company, didn't appreciate that he had defeated their tracker and fired him.

A GPS jammer on the ground in an urban area is a local nuisance. The jamming signals would get blocked by buildings before they got very far. But a 1-watt jammer on a balloon can deny GPS 100 kilometers away.

Here's a nightmare scenario for air traffic in the Northeast corridor. What if someone built a1-watt jammer, designed electronics so it came on for five minutes at random times, connected it to the car battery of a stolen car parked on the open roof of a high rise garage near Logan airport in Boston? How long do you think would it take before the jammer is located and disconnected? Answer: probably not before the car battery runs out.

NJ Turnpike runs by EWR. The arrow shows the location of GPS antennas for LAAS, an FAA system under development for GPS-guided approach and landing under poor visibility conditions.
(credit: Google Earth)

North Koreans apparently have switched on powerful GPS jammers aimed at their cousins in the south for 10 minutes at a time for days when they felt ignored.

Faking It: My GPS Says I wasn't There

Yes, you can rig things up so your GPS receiver shows you to be someplace else.

The motivation for cheating can be great. There are protected spawning grounds where fishermen are not allowed. If a fishing boat can sneak in, they can catch more valuable fish in one hour than they can do legally in a week. Think Papahānaumokuākea. The lawmen might be napping, but the GPS receiver required on the boat would give you away, unless you find away to 'fix' GPS.

Here's how you might do it.

Stop at the edge of the forbidden area. Rig up a system that receives and re-transmits GPS radio signals from this spot, jacked up a tad. We'd call this contraption a signal repeater. When you get the repeater working, switch off the GPS receiver on your boat for a couple of minutes and then switch it on again. Chances are the receiver would fall for the stronger signals from the repeater. How would you know? If the receiver continues to show the position of the floating platform you rigged up as you move the boat, you are all set. Go in and grab the loot. Your receiver would show you anchored in the legal area the entire time.

A much more challenging exercise would be to design a GPS spoofer which generates spurious GPS signals that make it appear to a tracker that a hijacked vehicle is following its original course. You can avoid paying road tax by making it appear that your car is on a parallel feeder road. An alert GPS receiver would catch on to such shenanigans and reject spurious radio signals, but there are a billion receivers out there that are too trusting and don't screen the signals as closely as they should.

Space Weather Hazards

A space event can wipe out GPS.

Solar storms can cause widespread power blackouts, disabling everything from radio communications to water supplies—most of which rely on electric pumps. They begin with an explosion on the sun's surface, known as a solar flare, sending X-rays and extreme UV radiation toward the earth at light speed. Energetic particles follow and these electrons and protons can electrify satellites and damage their electronics. Next come coronal mass ejections, billion-ton clouds of magnetized plasma that can take a day or more to cross the sun-earth divide. These are often deflected by earth's magnetic shield, but a direct hit may be devastating.

In 2011, the sun erupted with a powerful solar storm that just missed the earth but was big enough to "knock modern civilization back to the 18th century," according to NASA. A mass of swirling plasma rose up above the sun, twisted and turned for almost a day, then broke away. This extreme space weather was the most powerful in 150 years, but few earthlings had any idea what was going on. It could have been twice as bad as the 1989 solar storm that knocked out power across Quebec.

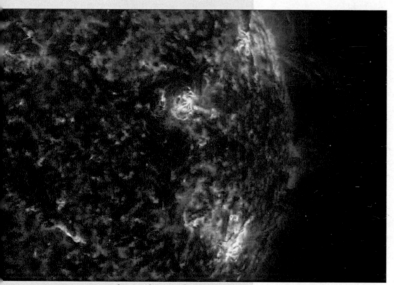

A mass of swirling plasma rose up above the sun, twisted and turned for almost a day, then broke away, Nov. 29-30, 2011. (NASA photo)

If the eruption had occurred only one week earlier, earth would have been in the line of fire. Instead the storm cloud hit the STEREO-A spacecraft, a solar observatory that is almost ideally equipped to measure the parameters of such an event. Scientists have analyzed the treasure trove of data it collected and concluded that it would have been comparable to the largest known space storm in 1859, known as the Carrington event, which had knocked out some of the telegraph systems of the day. A Carrington-size event could cost the modern economy more than a trillion dollars and cause damage that might take years to repair.

The Future: GPS & Wannabes

Success begets imitators.

And GPS, as you already know, has turned out to be an extraordinary success. But the first imitator appeared even before there were any such indications; others followed predictably in the 2000s when the GPS applications were growing explosively.

A GPS of Their Own

The Cold War logic dictated that the Soviet Union match the U.S. on every defensive or offensive initiative of any significance. So, the Soviets were just a couple of steps behind in starting development of a GPS of their own. It looked a lot like GPS. The system was called *Global'naya Navigatsionnaya Sputnikovaya Sistema* with acronym GLONASS (pronounced *gluh-naas* by the Russians, who took it over after the disintegration of the USSR, but **glow-nas** by others).

By the middle of the 1990s, there were signs that GPS was going to be big. Like anybody else paying attention, the Europeans recognized that this technology was vital and started making preemptive noises: GPS is owned by the U.S. military, and who can trust the U.S. military? It took a few years to build consensus among the member countries and the European Union launched in 2002 a program to develop a GPS of their own, to be called *Galileo*.

It also became clear in the 1990s that there is no stopping China in its march for economic and military preeminence in the 21st century. And an economic and military power needs a GPS of its own, and the Chinese called theirs *BeiDou*. That's Big Dipper in Chinese.

We refer to GLONASS, Galileo, and BeiDou as GPSs or GPS-Wannabes, but they actually have a proper, bureaucratic, generic name sanctioned by the U.N.: global navigation satellite systems (GNSSs). The definition of GNSS is malleable and a couple of regional systems under development in Japan and India are also covered under this rubric. GPS is a GNSS, and so are GLONASS, Galileo, and BeiDou. Japan's regional system QZSS and India's IRNSS are also GNSSs. The main difference is that GPS is used every day by a billion people and the others remain works in various stages of development. GPS has served as the model for all.

GLONASS, Galileo, and BeiDou didn't have to reinvent the wheel. Like GPS, all use trilateration and satellite constellations of about two dozen satellites in inclined medium earth orbits, and broadcast coded spread spectrum signals in the L-band. There are small differences in orbits and signal structures, but nothing basic.

A system designed in the pre-microprocessor, pre-Internet, pre-cellphone era of the 1970s would be expected to be obsolete by now. But not only is GPS not obsolete, it can't be beaten by the newer systems being designed from scratch in the 21st century. How does that work?

Why Do We Need So Many GPSs?

GLONASS-K

Galileo-FOC

BeiDou 2G

That depends on who *we* are—providers of service or customers. As we have said, the Soviet Union started developing GLONASS as a reflex action, but the motivations of Russia, Europe, and China are carefully thought out and similar: Basically, an economic or military power can't allow itself to become dependent on another for such a vital technology.

But it's not just about navigation. It's about sovereignty. It's about commerce. It's about high-tech jobs. It's about intellectual property and patents. And, of course, it's about military applications and national security.

But what about us, the civil users? Well, two GPSs are better than one, but not by much, for most of us. A 30-satellite GPS constellation gives us most of what we need. Where GPS doesn't work, the others wouldn't work either. But the GPS constellation can be cut back to 24, even 22, if the budgets get tight, and there is sense of security in having two autonomous systems—you know, no common-mode failure—especially if the signals are at different frequencies. Your receiver would use one system and have the other on standby. There is no point in wasting battery power by processing more signals than you need. The incremental value of a third system for you and me is near zero. A fourth one is worth still less.

Each system plans to provide one or more open signals to civil users. While the signals are free, the receivers are not. And, as we have said, it's all about economic activity and jobs and taxes. GPS, GLONASS, Galileo, and BeiDou would compete to build brand loyalty and sign up users. There is also the matter of national pride. Who is number 1?

Who is Number 1?

Of course, GPS right now is the one and only.

There is also no question that GPS has set high standards. When GLONASS, Galileo, and BeiDou take to the stage, GPS would have had 25 years to prove itself to a billion-plus daily users. It wouldn't be easy for the Wannabes to grab the spotlight. The competition among GLONASS, Galileo, and BeiDou is really for the number two spot. And, as noted above, *we* (the civil users) need no more than two systems.

Building a receiver that uses signals from multiple GPSs is not a challenge, but there is a cost associated with it. Why would a receiver manufacturer invest in developing a new receiver for GPS + X if nobody cares about X, or X may not be around in a few years.

We have already talked about GPS modernization, which will raise the number of open signals from one to four by 2020. If the modernization was an attempt by the U.S. to dissuade the Europeans from undertaking development of a GPS of their own, it didn't work. But we should be grateful that talk of a competing system gave us a more robust GPS.

But GPS can't afford to rest on its laurels. Ironically, a challenge for GPS lies in not falling victim to its own success. Getting funding in these times of belt tightening all around isn't easy. A billion dollars a year for operation and maintenance isn't going to be easy to find. GPS has to turn itself into a penny-pinching utility by finding ways to design cheaper satellites and saving money by launching a couple at a time.

What would it take for GPS to keep its position as number 1? The answer is that it must be reliable and dull like a utility. It must work like water supply and electric power. No drama.

Compatibility and Interoperability

Just look at the frequency allocations in the L-band for satellite navigation. There wasn't a lot going on in the L-band when GPS asked in the early 1970s for two 24-megahertz chunks. The band is now packed. As in any crowded situation, you want to make sure that the neighbors will act responsibly and not have loud parties. The idea is to tolerate each other and coexist peacefully. That's called compatibility.

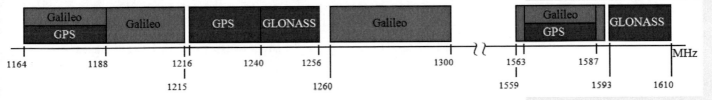

But there is no reason for GPS, GLONASS, Galileo, and BeiDou to stop at compatibility. They can take the next step and actually cooperate. No one's purpose is served if the position coordinates provided by GLONASS are different from those given by GPS, as they were until just a few years ago. Until now there was no reason for countries to agree on a common coordinate frame, or time standard. They will still maintain their own national standards, but determine how to transform the position coordinates from one coordinate frame to another and translate time to a different scale. The satellites will broadcast these parameters to be accounted for in a receiver.

The L-band was sparsely used in the 1970s when GPS asked for two chunks.

The Future of GPS Wannabes

We really should refrain from prognosticating about life with multiple GPSs. After all, prognostications about GPS by people a lot smarter proved so wrong. But we'll give it a shot.

Remember, development costs aside, it takes a billion dollars a year to operate and maintain a GPS-like system. For a Wannabe to be viable, the returns have to be commensurate.

GLONASS appears to owe a new lease on life to Putin's patronage and may have a less secure future beyond Vladimir Vladimirovich. There are only so many blows a system can withstand. GLONASS has suffered through the dissolution of the Soviet Union, destitution in the early days under Russia, and an extended moribund state at the start of the 21st century.

There was an embarrassing half-day break in GLONASS service in 2014. Then there was loss of three satellites in a launch in 2013 and a blow up in 2010 when three satellites tumbled into the Pacific shortly after launch. The system has a full satellite constellation in place and has been on the verge of usability for a few years, but when Putin goes, so could GLONASS.

Galileo began with brash assertions and cockiness: We'll show them how to build a satellite navigation system *right* and get rich off it, too! Galileo would be a joint public-private enterprise under civilian control, to be financed and managed by the European Commission, the European Space Agency, and European industry. It would cost €3.2 billion to develop and deploy the system, which would be ready in 2008.

The task, however, proved harder. Private industry showed no enthusiasm for investing in Galileo. The system is now planned to be operational in 2016 with public funds to the tune of €6.2 billion. The case for Galileo is largely economic. If Galileo can't secure a market and committed users, the *raison d'être* is gone.

China has so much momentum and a sense of predestination that we can't see them backing down on BeiDou. They have to project power. And they have an advantage of not having to explain themselves to anyone, or make a case for funds before balky representatives of the people predisposed to say 'no.'

There is no competition among the regional systems. Japanese, Indian, and Chinese regional systems have clear local benefits, are modest in their scope, and are likely to establish themselves as useful augmentations of GPS and BeiDou.

GPS at a Glance

What it is

■ GPS is a space-based radio-navigation and time distribution system developed and operated by the U.S. Department of Defense and paid for by the U.S. taxpayers.

■ The satellites broadcast precisely synchronized radio signals and a receiver simply 'listens' to them. There is no interaction with the satellites and the system can, therefore, support an unlimited number of users concurrently.

■ It's is a worldwide, all-time, all-weather service not affected by rain, snow, or fog. The only requirement is that the receiver antenna 'see' enough of the sky to receive signals directly from at least four satellites. The signals reaching the earth are very weak and any blockage can render them unusable.

How it Works

■ A GPS receiver measures ranges to satellites by determining how long it took a signal to get from each satellite to the receiver. The receiver also measures range rates (or Doppler frequency) for each satellite. Given the satellite position at the transmission time of the signal, the receiver calculates its position and velocity by solving a set of simultaneous equations exactly as we learned to do in high school algebra classes.

What it Provides

■ Position: Depending on the type of receiver, type of measurements made, and how they are processed, positioning accuracy can vary from a few meters to a few millimeters, permitting a wide range of applications from vehicle navigation to monitoring the movement of tectonic plates.

■ Velocity: Typically accurate to within 1 centimeter per second

■ Time: Typically accurate to within 100 nanoseconds

Satellite Constellation

■ 24-30 satellites arranged in six orbital planes with 12-hour orbits at an altitude of about 20,000 kilometers

Civil Signal (legacy signal transmitted in the open by all satellites)

■ Frequency: 1575.42 Megahertz (UHF band); Bandwidth: 2 Megahertz

■ It's free and open to all: no subscription, no license, and no fees

Timeline

- Mid-1960s: Concepts of GPS operation are developed.
- 1973: Program is approved by the Department of Defense.
- 1978: First prototype satellite is launched.
- 1989: First operational satellite is launched.
- 1995: System is declared operational.
- 2000: Purposeful degradation of the civil signal is discontinued and consumer applications take off.

Expenditure

- U.S. investment in design, development, and deployment: $ 20 billion
- Annual cost for operation and maintenance: $ 1 billion

Benefits

- Users: over 1 billion worldwide in 2015
- Most widely used U.S. military radio, albeit one way
- Civil receivers manufactured annually: > 500 million
- Annual commerce in GPS products & services: > $10 billion
- Economic benefits: $100 billion every year from increased productivity

Civil Applications

- Transportation: Land, air, maritime, and space
- Tracking: fleets of buses, trucks, taxis, ambulances, and police cars; children, pets, parolees
  - Industrial: surveying, construction, mining, agriculture, power grid
  - Commercial: time for financial transactions, digital networks
  - Scientific: geodesy and geophysics, weather, synchronizing observations at remote locations
- Personal: hiking, biking, running, fishing, golfing, crop circles, social media (where is everything and everybody I care about?), E-911

Information for GPS users

- For U.S. policy issues and general information: www.GPS.gov
- For GPS status: www.navcen.uscg.gov
- Acquisition Office: GPS Directorate, Los Angeles Air Force Base : www.losangeles.af.mil/library/factsheets/factsheet.asp?id=5311
- Trade magazines with lively coverage of the marketplace, technology, and policy issues: www.InsideGNSS.com, www.GPSWorld.com

Made in the USA
Middletown, DE
12 October 2016